THE WEATHER
AN INTRODUCTION
TO METEOROLOGY

Martin Frick

FOULIS
SPECTRUM BOOKS

A FOULIS SPECTRUM BOOK

THE WEATHER is published
by
The Haynes Publishing Group of Sparkford, Yeovil, Somerset BA22 7JJ,
England
First published in German under the title **WETTERKUNDE**
by
Hallwag AG of Bern, Switzerland

© English language edition, The Haynes Publishing Group 1976
First published October 1976

ISBN 0 85429 501 1

All rights reserved. No part of this book may be reproduced or transmitted in any form or by any means, electronic or mechanical, including photocopying, recording or by any information retrieval or storage system, without permission in writing from the copyright holder

English language translation Stanley Crawford for "Accurate Translations", Maidenhead, Berkshire
Colour photographs Martin Frick, Willi Gamper, MZA (Zurich, Switzerland), NASA, Heinz J Zumbuhl
Illustrations Monika Frick-Kutscha
English edition consultant Trevor Guymer
Printed and bound in Great Britain

CONTENTS

	Page
Why the weather should interest us	5
What is the weather?	7
The weather elements	12
How anticyclones and depressions are formed	21
Typical weather situations	26
Particular weather phenomena	36
The weather service and weather forecasting	57
Weather rules	79

WHY THE WEATHER SHOULD INTEREST US

Among the phenomena which man cannot avoid, the weather comes second in importance only to himself, and it is one of the most importunate things on earth. Nor is it only the extremes of weather, such as cold spells, heat waves and gales, or unusual events such as thunderstorms and hail, which compel our attention (all these can be a danger to life). No, it is above all the fact that every morning we have to test the weather with our senses, especially the temperature (and our interest in this matter is so superficial that we are no longer aware of it), if we do not want to risk feeling chilly all day because we have put on clothing which is too light or sweating because we are too warmly dressed. Weather is a dominant factor in our enviroment. It dictates fashion to an even greater extent than our fashion designers do, and whoever decides to disregard this usually has to pay for it with some sort of illness or other.

However, it is not only the happy-go-lucky person who has to come to terms with the weather; it applies even more so to the person who plans his activities, whether he be a simple hiker or an army officer. And it is precisely here that the whole background nature of the phenomenon of weather becomes apparent: it can be a source of vexation in one way, while in other ways it can be an agreeable companion, and this is why it has been the subject of many a fervent prayer.

The farmer is one of the oldest planners in human history. From time immemorial traditional rules have been passed on from father to son — when to plant this or that crop, the kind of weather in which that crop will thrive, and how and why the seed can be ruined. Because of this the farmer has been obliged to make careful observations of the weather for as long as anyone can remember. An occupation almost as important to that of the farmer in historical significance is the mariner's. The weather can be his best friend and provide him with favourable winds to carry him swiftly to his destination; it can, however, become a cruel foe, destroying both ship and life in a moment. This is how it has been up to now and how it will probably always be; at present man cannot control the forces in the atmosphere, but with increased observational awareness he can take some avoiding action to reduce loss of life and materials.

Weather has often intervened decisively in the course of history: many a victory and many a defeat can be ascribed to it. The commencement of the rainy season in tropical countries, and the arrival of winter in northern latitudes, must have terminated or checked many a campaign, coming as a blessing to peace-loving peoples. Weather is also a determining factor in the history of cultural development. A warm, sunny country like Italy will produce different customs, different art and different people from a country which is cold and has little sunshine. Politicians, psychologists, technicians, doctors and architects must all take it into account, while astronomers are always waiting for the sky to give them a clear view of the universe, just to give a few examples. Where there is no weather, away in outer space, man cannot live either, unless he surrounds himself with some monstrous technical equipment, and this is a condition which man could not really accept as 'life'.

The weather is as much a part of us as our body: it can heal or cause injury; it is with us always in all circumstances. It is, therefore, most important for people to learn about it and — to a certain extent — to get along with it. This is why, in every age, there have been people who have dedicated themselves to the subject of weather, and ever since the advance of scientific reasoning much loose speculation about the weather has given place to a genuine science of weather, to meteorology. Technology has provided weather science with a practicable basis; the introduction of the telegraph in the last century made it possible to observe weather events on a continental scale and thus to make some sort of start on a serviceable method of forecasting. In the present century, possibilities are opening out in a similar way following on the development of space travel and computer technology. Artificial earth satellites are photographing cloud patterns simultaneously over extensive areas of the earth's surface, thereby providing us with an overall picture from a viewpoint hitherto unknown or impossible of access, either to a person standing on the ground or even to one high up in an aircraft. Better satellite coverage should help in forecasting, particularly because it enables features to be identified in areas where conventional coverage is sparse. The data processing machines and computers are able in their turn to master data in quantities which the brain can no longer handle and with which man, for all his versatility, would not like to burden it; how pleasant it is to be able finally to delegate to machines the boring and monotonous jobs which our practical sciences have brought with them!

The aim of this book is to provide an introduction to meteorology and its methods, paying particular attention to the weather of the northern temperate latitudes. Other regions of the world have climatic characteristics of their own.

WHAT IS WEATHER?

Weather is the name we give to those visible effects taking place in the envelope of air, or *atmosphere,* which surrounds our earth. Without it there would be neither wind nor rain, neither clouds nor thunderstorms, no hail, no snow and no dew. No mists would hang in the valleys between the mountains, no rainbows would arch overhead and we should never see the rosy glow of a sunset sky.

In all its manifestations the weather is bound up with the atmosphere, or rather directly with the power part of it, called the *troposhere.* The height of the upper boundary of the troposhpere, the *tropopause,* varies between 5 and 8 kilometres in the region of the poles, and 12 to 15 km at the equator. It is distinguished from the stratosphere above it — in the troposhere there is a general decrease of temperature with height, whereas in the lower stratosphere the temperature is nearly constant.

It is very likely that the weather processes mentioned above are, long term, part influenced by those taking place in the stratosphere and above, which can be shown to extend to at least 600 kilometres above the earth's surface; they gradually merge into the vacuum of outer space. The thing that is important for us, that we can see and feel, and that we call "weather", this all takes place in the layers immediately above the solid earth's crust and immediately over the oceans.

What is air? It is a mixture of gases which also contains a certain amount of dust. It is composed mainly of two gases: nitrogen and oxygen, which together make up about 99 percent of the air. In addition to these, there are the important constituents, carbon dioxide and water vapour. Finally, there are very tiny quantities of a number of rare gases, the most important being argon, helium, krypton and neon. Up to a height of 20 to 30 kilometres, the ratio of nitrogen to oxygen is constant at about 4 : 1; this means that the ratio of the quantities of the gases is always the same. The absolute amount of each is, of course, very different.

As we have said, air is a gaseous mixture, or more briefly, simply a gas and, like all gases, it has a propensity to expand. The only direction in which the air can do this is into outer space. Hence we have the tendency of the atmosphere to expand as far away from the earth as possible. If it were not for the force of gravity exerted by the earth, all the particles of air would fly off into space.

There is another thing to be said about the atmosphere, if we are to get a rather clearer idea about how weather phenomena of every kind are produced. We said that air consists principally of nitrogen, oxygen, water vapour and carbon dioxide. The last-named has a less noticeable, but extremely important, part to play — it is a heat insulator for the earth. It lets the sun's rays come in but prevents the long wave radiation (the infra-red) emitted by the earth, from going out. Solar radiation, is of course, in the short wave part of the spectrum.

Our globe receives energy from the sun in the form of heat (invisible, long wave radiation), light (visible radiation) and ultra-violet (invisible, short wave radiation). That part of radiation which gets through the atmosphere and reaches the earth's surface is absorbed there. It warms up the ground which, in turn, emits long wave radiation into the atmosphere; this continuous two-way radiation causes the atmosphere as a whole to suffer long wave cooling and thereby causes the evaporation of water. In so far as it warms up the air, it directly performs work: rising currents of air, rarefaction, wind. the actual amount of heat lost in this way depends on the state of the atmosphere. This is the reason for the rapid and marked cooling in mountainous regions, where the air is thin and pure.

The weather phenomena we see — the clouds, snow, fog etc — are restricted to the lower atmosphere, that is the troposphere. Weather occurs in the tropoposphere because of a complicated balance between radiation and convection processes. In order to complete the energy balance warm air from the equator must be transferred polewards and slightly upwards, being replaced by cold air from higher latitudes The air needs to go no higher than 10 km to lose sufficient heat by radiational cooling. There are always large amounts of solid particles of all kinds suspended in the air; dust which has been blown up from the ground, smoke, volcanic ash and many other things. Because of their weight and the stability of the air they occur far more frequently near the ground than higher up. While the air at the height of the alpine peaks is almost dust-free and sterile, that in and over our cities contains many millions of particles per cubic metre. Pollution is also a contributory factor determining visibility. This decreases rapidly with increasing humidity because there is a considerable increase in the size of the dust particles due to water vapour attaching itself to them. After rain has washed out the air it becomes clear and pure again. The saying "Oh sky so blue from days so dull" is not without foundation. Besides this kind of dullness, there is yet another, the so-called 'turbid light' or 'aerial turbidity'. It has two causes: first, the action of very fine, irregularly heated filaments of air ('convection currents' on a small scale) — they are responsible for the familiar striated shimmer of the air over ground irradiated by the sun; the second is the scattering of light by the air molecules themselves. It is this scattering which causes the colour of the air. Blue light is much more strongly scattered than red, and therefore there is a preponderance of the blue colour which is, in itself, colourless. This only applies to scattering by the very small particles, the size of which is less than the wave length of the light. Larger particles act according to the laws of geometric optics, and are neutral in their reflection of white light. The more dust, water droplets, etc., there are floating about in the air, the more the white light will predominate. We have already suggested that the air is a mixture of various gases and that its composition is everywhere the same (even up to great heights). This applies to dry air. However, the air is never completely dry; it always contains water in a gaseous form, that is water vapour.

While all other gases are present everywhere in the same relative proportions, the water vapour content shows marked variations.

This water vapour is much lighter than dry air. Air pressure which is indicated by our measuring instruments is, however, the sum of the partial pressures of the individual constituents of the air.

At a constant temperature, a given volume of air can only absorb a certain maximum amount of water vapour. Once this value is reached, we say that the air is saturated. The quantity increases with increasing temperature, though by no means at a constant rate.

The further the air is from saturation, the more rapidly it will take up water from a moist object and the more rapidly the saturated layers of air surrounding the object are removed, to be replaced by new unsaturated layers. We can now see why extremely warm, dry winds have such an extraordinarily 'parching' effect and make us so thirsty.

The more space available to a given quantity of water vapour, the more the latter expands and the more water vapour can be taken up before saturation is reached. In other words, expansion of the air at constant temperature increases its capacity for absorbing water. Hence, rarefied air enhances evaporation and dries things out more rapidly. The evaporation increases at an extraordinary rate with expansion of the air or with increase in the height to which we ascend (providing, it must be repeated, that the temperature remains the same). This does not entirely account for the great thirst which people experience in the mountains, for it is also felt when it is raining or in low-lying country when there is fog or snow. It has more to do with the 'physiological saturation deficit'. When we breathe in air which is saturated at, say, zero degrees or below, it is heated up in our respiratory organs almost to body temperature while its water content remains the same. The amount of water which it could hold at body temperature is much greater; it is far from being saturated and therefore the water from out membranes can be evaporated into it at a rapid rate. Our body reacts to this with an intense sensation of thirst.

In scientific terms, it can be expressed like this: with increasing height the pressure of the air diminishes more slowly than the vapour pressure. This is why every trace of perspiration vanishes so quickly (as occurs on the high peaks in the Alps) and why snow disappears without melting.

If the air saturated with water vapour is cooled, then the water begins to separate out in liquid form. The details of the process will be examined later. Here, our primary concern is the great importance of water in the air for the balance of nature and for explaining the creation and shaping of the weather. The oceans are kept relatively warm by the sun — and they cover seven tenths of the globe — and the flow of cool dry air across them raises up enormous quantities of water by evaporation. All this huge amount of water is carried by the air currents to other places; this cannot occur without at the same time taking along colossal quantities of heat (an extremely important point!) for evaporation results from the action of solar energy, the work performed being incorporated in the vapour in a dormant form. Once the vapour condenses into liquid or solid form (ie; cloud) the heat is liberated again. The moist air masses which come to us from the ocean affect us, then, in numerous ways.

First of all, they bring along that heat which corresponds to their temperature and specific heat and, in addition, the energy which is bound up with the accompanying water vapour. The moment that cloud begins to form, warming of the air mass occurs. Now all the energy, which had been taken up over the ocean along with the evaporated sea water, is liberated once more as the water vapour condenses to form clouds. If the rain reaches the ground there is cooling because much of the water evaporates again, while the remainder flows back to the sea in streams and rivers. The cycle is complete. This explanation is, of course, an over simplification, for it avoids the much more complex process of the depression systems. It does, however, illustrate the basic system.

The subjective feeling that rain has 'broken' a cold spell usually arises, either because one finds oneself in cloud (in the mountains, for example), or because the wind falls off with the onset of precipitation and this is mistakenly interpreted as 'warming up'.

The humidity of the air affects us in all kinds of ways which are mainly beneficial. Without it, there would be no rain, or snow, and so the soil would not be fertile, the air would not be cleansed of its impurities and the moderating effect on temperature would be much less. We should then have an extreme desert-like climate, with an intense emission of radiation to outer space from all those surfaces which had received radiant energy from the sun.

That intricate physical system, the troposphere, is the theatre in which all those phenomena which make up what we call the weather take place — complicated physical processess, interconnected in a way which is difficult to discern. Above the troposphere, there are further atmospheric layers which influence events in the troposphere to a greater or lesser extent (see Fig. 1). Within the stratosphere, which can reach a height of about 60

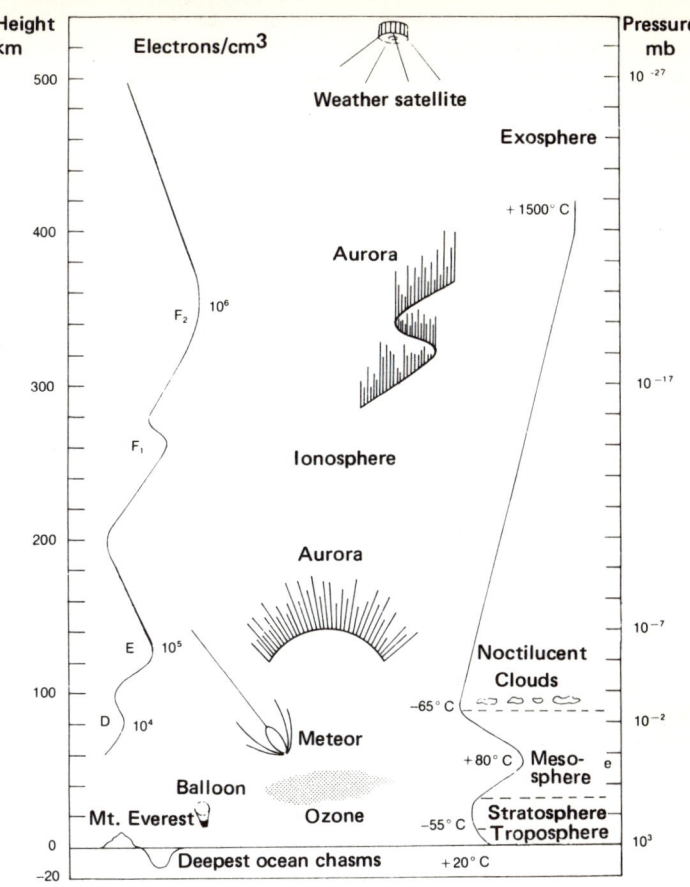

Fig. 1 Diagram of the earth's atmosphere. Air pressure decreases with increasing altitude

kilometres, there is the ozone layer with its maximum concentration at about 30 kilometres. This layer is extremely important as far as life on earth is concerned; without it no life would have been able to develop, at least not on land. This layer of ozone, consisting of triatomic oxygen, absorbs the dangerous ultra-violet radiation emitted by the sun; strictly speaking, not all of it! It is the skier's experience that he becomes tanned in the mountains, even if he is in cloud. This is due to the small amounts of UV-radiation which get through the ozone layer and which we, down at sea level, only perceive to a lesser degree because of its further attenuation in the lower layers of the atmosphere.

Above the stratosphere there are other layers which are more important in the science of geophysics than in meteorology. The principal one is the ionosphere which affects radio propagation. Its electric charge, which is generated by particle and UV-radiation from the sun, is of decisive importance for the radio waves which are reflected from it back to the earth. Weather forecasts are put out for the ionosphere, just as for the troposphere, but obviously they are primarily of interest to people concerned with radio technology.

THE WEATHER ELEMENTS

If we want to get an idea of the complicated things going on in our atmosphere, and do it in a scientific way, then we must keep to measurable characteristics of the weather. Observing the weather as a matter of measuring characteristic physical quantities correctly is just one aspect of a meteorologist's work. Following this, there has to be an examination of all the measurements so diligently collected in order to disclose any recognizable laws involved in the mysterious events. The meteorologist is not, however, content with this alone. He is far more interested in the reverse process and the advantage which might be gained from it, viz, the possibility of using recognized laws and further observations to forecast future events. The success of such forecasts is, first and foremost, the confirmation of theory. It is here, however, that there remains much to be desired. The flaws in weather forecasts are evidence of the complexity of weather phenomena and how difficult they are to comprehend.

Well, what are the weather elements we have to observe and measure?

In attempting to reduce the weather to as small a number of measurable physical quantities as possible, it has been found that the temperature, pressure and humidity of the air are the most suitable ones for this purpose. Combined with these simple elements we have three others: wind, cloud and precipitation. These six elements do not, of course, represent the only factors which have to be taken into account in weather forecasting. This is the point at which we begin to speak, rather unphysically, of 'air masses'. When a long, quiet period of fine summer weather comes to an end, with cloud increasing and rain setting in within a few hours, this is due to another air mass invading our area and displacing the air which had been there previously. It has different physical properties, and these result in a change of weather. There are four main types of air mass: equatorial, subtropical, that originating in temperate latitudes, and arctic air masses. These are further divided into two sub-groups: continental and maritime, according to whether they have originated over the land or over the ocean. Air which has remained for a long period over land brings with it a lower humidity than air which comes to us from the sea.

Air temperature

Temperature is one of those quantities to which people are directly sensitive but, like any sensation, it is subjective, and therefore can often mislead one into estimating its value incorrectly, ie giving it a value with which not everybody agrees. Even with a thermometer, one can frequently give the temperature incorrectly. The measurement of temperature is almost a science in itself. When Aunt Joanna declares that the mercury was up today, she probably says nothing about her thermometer having been in the sun all day or that it is mounted directly onto a sheet of metal on her balcony. She might just as well have put it in the frying pan, where it would have become even hotter. What we really want to measure is not the temperature of an object which has been heated by the sun's radiation, nor the temperature of the air close to a wall or over an asphalt road where it can be heated directly by convection, but the temperature of the air in the open. This is the only temperature which is characteristic of the weather. Meteorologists therefore mount the housing, or meteorological screen for their instruments, over grass, at a definite height above the ground and painted white, since a bright object reflects most of the radiation, while a dark one absorbs it and converts it into heat. The thermometer and other instruments are mounted inside the housing.

Now a word about the units of measurements. When we read in the newspaper that New York has had a summer 'high' temperature of 110 degrees, we would not be unduly surprised. If the reporters had converted the Fahrenheit degrees used in America into our (now official) Celsius (centigrade) degrees, we might have been. The Celsius scale is arranged so that $0^{\circ}C$ corresponds to the melting point of ice and $100^{\circ}C$ corresponds to the boiling point of water (at normal air pressure). The thermometer contains a thread of mercury or some other liquid, the length of which varies with the temperature. Thermometers can also make use of the fact that certain substances change their electrical resistance with change in temperature.

Air pressure

Man possesses no special organ for air pressure, although the latter acts in conjunction with other factors to affect his state of health. From the physical point of view, air pressure is quite easy to measure with a barometer. We can use the height of a column of mercury, the weight of which is the same as that of the column of air over the place of observation; or we can use an evacuated box made of flexible metal which the pressure of the air compresses to a greater or lesser extent. The units employed in measuring air pressure are the torr (or mm of mercury, ie, the height of the column of mercury mentioned above) or the millibar. 1 mb is equivalent to 1000 dynes per cm^2 of surface; 1 dyne is the force which produces an acceleration of $1cm/sec^2$ in a mass of 1 gram. 1000 mb correspond to 750 torr. The 'normal' or 'standard' air pressure of 1013 mb or 760 torr corresponds to a physical atmosphere (1 atm). If we want to compare the air pressure at one place with that at another place, the readings of the two columns of mercury should be reduced to a common temperature, for example, $0^{\circ}C$ (the standard temperature). There is another point to consider. To make these comparative readings of air pressure using the barometer it is necessary to compensate for the temperature. Two such adjustments have to be made. The first is to correct for expansion in the instrument and give the air pressure at station level, which in general is not at mean sea level. To reduce to MSL a correction also has to be made for the imaginary column of air which would exist below the barometer. The pressure difference between the tap and bottom of this column depends on the temperature. A reasonable working approximation for most temperatures encountered near the surface is 1 mb to be added for each 8 m above MSL. This has nothing to do with the weather, but is simply due to the fact that, as we ascend, the column of air which stretches upwards above us right to the upper boundary of the atmosphere becomes significantly shorter — 'significantly', because the lower part of the atmosphere is much denser than the upper part. A barometer can, therefore, also be used to measure altitude.

The fact that air pressure is not the same at all places is of the greatest importance as far as weather is concerned. The smallest pressure difference between two places causes air to move from the place with the higher pressure towards the place with the lower pressure. In other words, a wind is created. The larger the pressure difference between the places in relation to their distance apart, the more quickly the air will flow down this 'pressure gradient' (the quantity 'pressure difference divided by distance apart'). There are, of course, other forces acting on the air such as those associated with the rotation of the earth and friction at the earth's surface. So much for the physical fundamentals. Later on, when we discuss high and low pressure formations, and the wind, we shall have more to say about air pressure.

Humidity

This quantity has a particularly marked effect on man, but is not so easy to measure. As already explained, air is able to absorb a certain amount of water vapour — cold air less than warm air. If we express the maximum amount of water vapour absorbed in grams per cubic metre of air, we have the following figures:

Air temperature	$0°C$	$10°C$	$20°C$	$30°C$
Approximate maximum water vapour absorbed per m^3 of air	5g	9g	17g	30g

Air which has a temperature of $10°C$ and contains $9g/m^3$ of water vapour is saturated. What happens if the temperature rises? At the higher temperature, the air can absorb more water vapour and is, therefore, no longer saturated. On the other hand, if the temperature goes down the air cannot hold as much water vapour, and the excess amount which it contains separates out in the form of droplets. This is how fog and clouds are formed, and, eventually, rain and snow. The humidity, expressed in g/kg, is called the specific humidity. Since the effects of humidity depend primarily on how much moisture the air actually contains, relative humidity has been introduced. It is expressed as a percentage. If we call the specific humidity, ie the actual water vapour content of the air, q, and the maximum amount which it can absorb at the prevailing temperature qs, then the relative humidity rh is given by

$$rh = \frac{q}{q_s} 100\%$$

When air rises in the atmosphere, it also cools, owing to the expansion reaches the point at which its specific humidity, q, equals the maximum value, qs, and the relative humidity is thus 100%. At this stage, moisture, separates out, and we can see why this temperature is called the dew point. The thermodynamics of humid air is a fairly complicated business. In order to understand certain weather phenomena, we need to know how moist air behaves at the higher levers of the atmosphere. When air is compressed, it becomes warmer, as can easily be demonstrated with a bicycle pump. If the tap of a cylinder of carbon dioxide is turned on, the stream of gas which comes out is freezing, the sudden expansion having caused intense cooling. When air rises in the atmosphere, it also cools owing to the expansion which takes place as a result of the decrease in air pressure with height. As long as no condensation sets in, the cooling amounts to 1 degree per hundred metres.

As cooling proceeds, the relative humidity increases, because the capacity of the air to absorb water vapour, ie the maximum humidity q_s, decreases. Once the dew point is reached, the water vapour in the air begins to condense out as droplets. The process is accompanied by the liberation of heat which raises the temperature of the air. As a result, any further upward motion of the air reduces the cooling to about ½ degree per 100 metres. We shall return to this later.

Measuring humidity is more difficult than measuring pressure. Hair hygrometers are used. In these instruments, the change in length of a bundle of human hair with changes in humidity are converted by a lever mechanism into a pointer reading. A thermometer, the bulb of which is wrapped in a moistened piece of material is also often used, although for upper air measurements a wet bulb thermometer cannot be used because it freezes early in the ascent as temperatures fall below $0^{\circ}C$ and fails to give the true relative humidity. When the ambient air has 100% relative humidity, no evaporation takes place from the moisture in the fabric, and the thermometer shows the same temperature as it would in the dry state. If the air is not saturated, however, moisture evaporates from the fabric, and the more so, the drier the air; also the more the evaporation from the material, the more strongly the thermometer is cooled. A measurement of the cooling can be used with the appropriate tables to find the value of the humidity. One can also use the electrical resistance of certain substances which attract moisture (hygroscopic substances) to obtain a measurement of the humidity of the air.

Wind, cloud and precipitation

Precipitation is measured in a rain guage. This consists of a large funnel with a measuring cylinder underneath for reading off the amount of precipitation in mm (an allowance is incorporated for the area of the funnel). Measuring the cloud amount is less satisfactory. One has to be content with estimating the fraction of the sky which is covered. An indirect measurement can also be obtained by using a glass sphere, which acts like a burning glass (convex lens) and has a strip of card fixed behind it. As the sun moves acorss the sky during the course of the day, it burns a trace on the card; any interruptions in this trace are ascribed to cloud. For weather statistics, the numerical values of precipitation and cloud cover are, of course, very important. For weather forecasting, however, the type of cloud and precipitation is much more important, but these are characteristics which cannot be expressed in numerical values. What is required is a detailed list of the individual types, and this will be provided. First, however, let us see what other basic physical information we require.

There are a few more points about the wind. The wind is a directed quantity, or vector, and so we have to measure its direction as well as its force or speed. From the earliest times there have been some quite amusing devices invented for this purpose. It would hardly be possible to assign a date to the invention of the wind vane. For wind speed we have the rotating cup anemometer; this incorporates a generator and produces a voltage which depends on the rate of rotation and thus indicates the wind speed. There are also precision instruments which count the number of revolutions of the cups and so give the run of the wind over a given interval of time. For estimating the strength of the wind we still use, even for scientific purposes, the Beaufort Scale of wind force, named after the British Admiral, Beaufort, who introduced it in 1805. On this scale, the wind force is assigned a numerical value according to its visible effect, although Beaufort originally expressed the scale in terms of the sail carried by a man-of-war!

The matter of wind direction, however, is rather more complicated. As long as we are not considering just a local wind, but one which is blowing along a sufficiently long track, then in reality it does not simply flow down the horizontal pressure gradient. There is a deviation in direction due to the earth's rotation. In the northern hemisphere, winds are deflected to the right and in the southern hemisphere to the left. We need only consider the radial component of the vector which represents the earth's angular velocity at the place in question (see Fig. 2). It can then be seen that, in the northern hemisphere, the earth is turning under the wind in an anticlockwise direction. It thus seems to us — at rest on the earth's surface — that the wind is turning in a clockwise direction. This process is generally described by invoking the principle of the so-called 'Coriolis force'. Fig. 3 shows that between a region of high pressure, a 'high', and a region of low pressure, a 'low', there exists a pressure gradient. The continuous lines are lines of equal pressure (isobars). The pressure gradient at every point is perpendicular to the isobars, but the wind direction is turned quite a bit to

Beaufort Scale   Average wind speed (to the nearest whole number)

	km/h	knots
0 *Calm or almost calm* Smoke rises almost vertically. Sea like a mirror	1	1
1 *Light air* Wind direction shown by drift of smoke but flags hardly disturbed. Ripples appear here and there	3	2
2 *Light breeze* Wind felt on face, leaves move on trees, flags partly unfurl. Sea slightly ruffled by widespread ripples	9	5
3 *Gentle breeze* Leaves and small twigs in constant motion. Sea gently undulating	16	9
4 *Moderate breeze* Raises dust and bits of paper, small branches move, flags completely unfurled. Moderate waves, white horses here and there on larger stretches of water, surface very rough	24	13
5 *Fresh breeze* Small trees in foliage and larger branches without leaves begin to sway, flags fully extended. Isolated white horses on small stretches of water, several white horses on larger stretches of water	33	18
6 *Strong breeze* Large branches in motion, whistling heard in telegraph wires, wind can be heard against houses and fixed objects. Crests drawn out into spindrift on larger stretches of water, occasional spray when waves break, several white horses on smaller stretches of water	45	24

	Average wind speed (to the nearest whole number)	
	km/h	knots

7 *Near gale*
Whole trees in motion, resistance felt when walking against wind. Here and there, on larger stretches of water, foam blown in streaks parallel to the direction of the wind — 56 — 30

8 *Gale*
Breaks twigs and small branches of trees, walking against wind more difficult. Over large stretches of water, foam frequently blown out in streaks along in the direction of the wind, water blown from the crests of waves, occasional spray now blown in streaks over smaller stretches of water — 68 — 37

9 *Strong gale*
Large leafless branches broken off, minor damage to houses (roof tiles and chimney pots removed). Widespread streaks of foam and spray, visibility reduced in some directions — 82 — 44

10 *Storm*
Whole trees broken off or uprooted, serious damage to houses. Over large stretches of water, extensive areas of foam blown in dense white streaks, sea takes on a generally white appearance, visibility reduced in all directions — 96 — 52

11 *Violent storm*
Widespread damage, very seldom experienced inland — 110 — 60

12 *Hurricane*
Widespread devastation — 110 and over — 60

The strength of the wind obviously depends on the pressure difference in the atmosphere. The bigger the pressure difference and the smaller the associated distance (that is, the larger the horizontal pressure gradient), the stronger the wind blows.

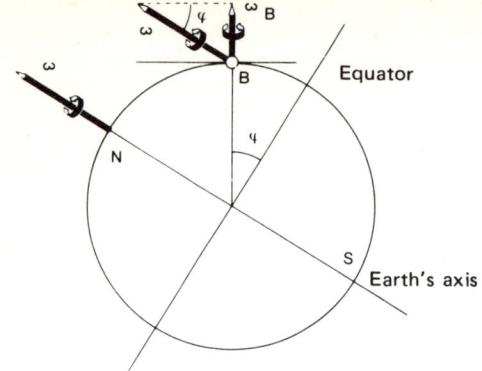

Fig. 2 Diagram of the deflection of the wind in the northern hemisphere due to the rotation of the earth

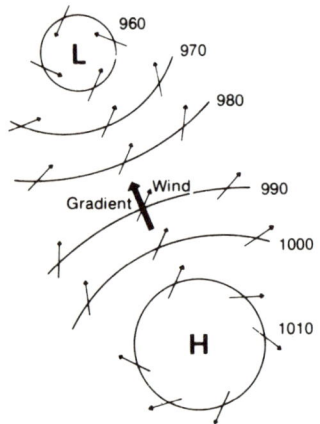

Fig. 3 Wind direction between high pressure (anticyclone) and low pressure (depression)

the right of it (in a clockwise direction). Seen as a whole, both the high and the low have a spiral of winds, but the spirals are winding in opposite directions. The air is flowing out of the high in a clockwise direction and into the low in an anticlockwise direction. These two formations are also called cyclones or depressions (low pressure areas) and anticyclones (high pressure areas).

The deviation caused by the earth's rotation is so great that the winds have the tendency even to blow parallel to the isobars, the upper winds actually do, but only if the isobars are straight and equally spaced. At ground level, however, there is friction between the air and the earth's surface, as a result of which the surface wind blows at an acute angle across the isobars. The effect can best be remembered in this way: if we stand with our backs to the wind (in the northern hemisphere), then the low pressure is on the left and a little in front, while the high pressure is on the right and a little to the rear. Historically this is known as Buys-Ballor's Law after the Dutch professor.

HOW ANTICYCLONES AND DEPRESSIONS ARE FORMED

Simply, the reason why pressure differences arise is always to be found in the unequal heating of the earth's surface. Over a warm place, the density of the air and the air pressure are lower because the air there has expanded; at a cold place, the density and pressure are greater. As far as the motion of the air is concerned, the results can be observed locally over the smallest area as well as on a global scale. Even the different heating of field and forest, of slope and plain, causes a wind. The glider pilot is very familiar with this. The conditions on the coast are very clearly marked. During the day, the land warms up more quickly than the water and the pressure falls over the land. Over the sea, the pressure is high and a sea breeze sets in (wind from the sea towards the land). The air over the low pressure moves upwards and flows out in the direction of the sea at some higher level. The circulation which is created in this way falls calm towards evening, and reverses its direction during the night. This is because the water retains the radiant heat it has received longer than the land; the lower pressure is now over the sea and a land breeze sets in.

The same things happen on a large scale with the oceans and the continents, though in this case the deflection due to the earth's rotation has to be taken into account. Again as an oversimplification, as the warm moist air from the subtropics slides past the cold dry air from the polar regions, waves are formed which have some resemblance to those on the surface of water over which the wind is blowing. As the amplitude of the wave develops, a depression is formed (see Fig. 4.).

Fig. 4 Diagram showing the formation of waves between warm and cold air masses which are sliding past each other. As a wave becomes distorted, a depression is formed

Weather phenomena in the depression

During the formation of the vortex on the surface of separation, there is no immediate intermixing of the two completely different air masses (fronts); in fact, the boundaries become sharper. On the contrary, the most varied weather processes take place at the boundary surface as shown in Fig. 5.

Where the warm air at the surface gradually displaces the cold air, we have the warm front. The warm air, being lighter, also slides up over the cold air. The cold air to the rear of the depression thrusts, moving faster, underneath the warm air of the warm sector (at this position we have the cold front) lifting it away from the surface, and this can eventually lead to there being only a single disturbance line left (called an occlusion).

As the warm air slides upwards it cools. The considerable quantity of water vapour, which it was able to store because of its higher temperature, now gradually separates out. This produces very uniform layers of cloud along the boundary surface. In summer, the turbulence within the cloud is set up because of latent heat release which causes small-scale instability. The physics of clouds is complicated, but a simple understanding may be gained. As the air continues to ascend after reaching saturation, the cloud layer becomes sufficiently deep for cloud droplets to develop, having different sizes. The larger the drop, the faster it will fall through the air, colliding with other drops as it does so. Eventually drops which are large enough to fall to the earth's surface will be formed and these will drop from the bottom of the cloud and some may reach the ground before evaporating. In parts, the vertical air motion is small and so the drops do not have to

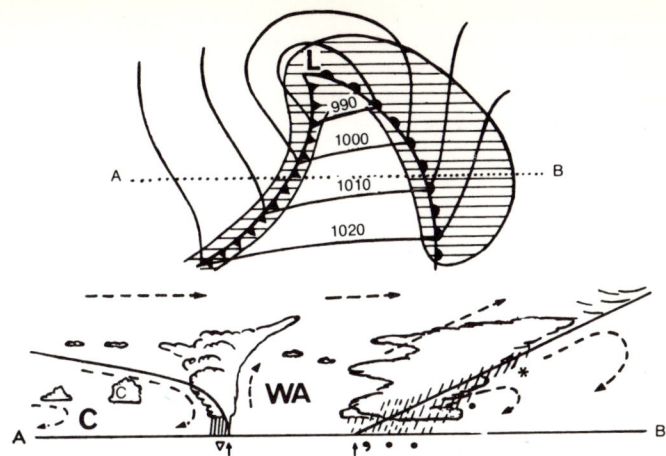

Fig. 5 Diagram of the weather phenomena which can take place at the boundary surface between different air masses

▲▲	warm front	•	rain
▲▲	cold front	*	snow
1010	lines of equal pressure (isobars)	?	drizzle
		▽	shower
L	centre of the region of low pressure	↑	warm front or cold front at surface
≡≡≡	area of precipitation	WA	warm air or warm sector
....	cross section A—B	C	cold air or rear (of depression)
⟶	stream-lines in the warm and cold air		

grow to a very large size before they fall out. In showers the air in some places ascends at several metres per second and only when the drops are quite large will their fallspeed be great enough for them to reach the ground. Hence the difference in character between frontal and shower-type rainfall.

The passage of the warm front can be recognized by the temperature rising gradually and the precipitation gradually ceasing. Warm sectors in Britain, particularly close to the centre of a depression, are quite often completely cloudy with drizzle or light rain.

Precipitation at the cold front can very considerably in character. Sometimes the rain lasts for several hours with only a gradual clearance. On other occasions showers become organised at the surface front and large changes of wind and temperature occur with a dramatic clearance afterwards.

This is where the isobar picture gives us some important information. As we have already seen, the upper wind blows more or less parallel to the isobars, and the greater the pressure gradient (ie the closer the isobars are to each other), the stronger the winds. If we want to know which direction the disturbance is coming from, we must assess the direction of the low-level wind and the movement of the high clouds (cirrus-level wind) running ahead and choose the direction somewhere between the two.

The 'intensity of a depression' (ie the air pressure at its centre) is, of course, decisive for the strength of the winds in the vortex. Intense lows such as hurricanes or typhoons, tend not to occur in our latitudes, while depressions seldom bring us severely devastating storms. Our depressions are mostly less than deep but, on the other hand, very extensive. In this connection, it is not the barometer reading which is so important for weather forecasting, but the rate at which it is rising or falling. If a deep depression is approaching, the pressure will fall more quickly than in advance of the warm front of a gentler vortex; at the cold front it rises abruptly. Unfortunately, this simplified representation of the fronts cannot be immediately transferred to the region of the Alps (see next paragraph). If the depression is approaching from the south-west, the isobars ahead of the warm front run almost south to north. This gives a wind transverse to the mountain range, which results in a fohn. This causes the warm front to be obliterated or to dissolve, and it becomes ineffective in central Switzerland. Precipitations from the warm front fall in western Switzerland, and over the Jura and its northern extension. One can even say that it is only in rare cases that the warm sector is not associated with fohn conditions. In areas which are subject to it, there is a quite considerable rise in temperature.

The Fohn *(a European phenomena associated with Switzerland)*

This is an unpleasant intermediate state between fine weather and poor weather, and is a terror to many weather-sensitive people. 'Fohn' was not originally a German word. It comes from the Latin *favonius*. This was the term used for a very pleasant mild westerly wind, but the Roman conquerors, now on the northern side of the Alps, also gave the name to the warm, dry winds which came down from the mountains as southerlies. Let us imagine that the wind is obstructed by a mountain range, 1,000 metres high. Suppose the air in it is dry. Then, as we have seen, ascending the windward side of the mountain it will cool 1 degree every 100 metres, arriving at the summit with a temperature 10 degrees lower than it had been at the foot of the mountain. As it descends on the other side, the air will warm up again by 10 degrees. In the case of the fohn, however, the ascending air is moist, and it will cool and reach saturation at a certain height. Clouds will form and then precipitation will occur, and during further ascent the cooling will be only ½ degree per 100 metres (see Fig. 6). As an example, suppose the air has to ascend 400 metres to reach saturation; then, by the top of the mountain, the total cooling will have been only 7 degrees! When the air descends through 1000 metres on the other side, it will warm up, as in the first example, by 10 degrees. Thus, the wind on the leeward side of the mountain will be 3 degrees warmer than on the windward side. Fohn air is clear and brings unusually good visibility; the weather is warm and dry. The only trouble with it is that pressure is low with a tendency to fall, and this is what, it is thought, affects sensitive people, although research is still going on.

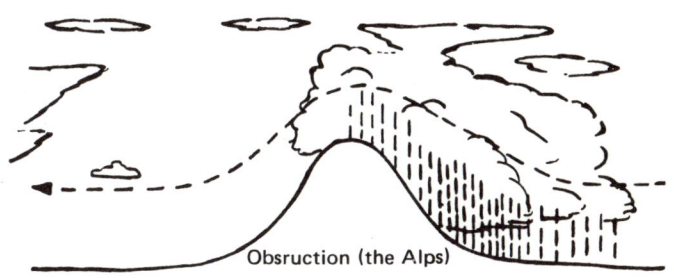

Fig. 6 Diagram of a 'fohn' situation

Frequently, the fohn blows into an approaching depression and distorts its normal weather sequence. This continues until the effectiveness of the fohn is brought to an end by its mixing with colder air. The barometer then rises, but poor weather, with cold and rain sets in! It is hard to forecast when the fohn is going to 'collapse'. Fohn situations are rather difficult to work out, and they give the weather prophets more trouble than other weather situations.

TYPICAL WEATHER SITUATIONS

It is only possible to chance a forecast for a longish period (several days), when a weather situation can be recognized as having some semblance of durability. This means that a stable anticyclone must have formed in the general Atlantic — European area. According to the position of the anti-cyclone and the relative disposition of the depressions, we can distinguish various characteristic types of weather. Some of these will now be explained. The weather charts are only schematic.

The tracks of depressions

Low pressure areas are in the habit of coming along in families. Often a couple follow one another on the same track. These are not just any tracks. On the contrary, certain ones are favoured, and this is an important point in weather forecasting. In Switzerland, between one depression and the next of the pair, the weather usually clears up temporarily. Behind the 'final' cold front, however, we get a *Stau* situation along the Alps (a damming up of the air by the mountains), which produces very poor weather for 1 to 3 days (the further we go into the mountains, the longer it remains around). The sky is uniformly cloudy, just as before a warm front, but the precipitation is of varying intensity. There is a rule of thumb which warns us that the quicker the clearance behind a cold front in the mountains, the briefer the following fine weather will be. The *Stau* situation is brought to an end by the *Bise* (the local name for a certain type of north to north-east wind). In western Switzerland, it is called the *Bise noire* because it carries back the *Stau* cloud which had been swept away to the east, thus bringing the interval of bright weather to a sudden end. In winter, of course, in the broad valleys of the central region of the country, it is not long before fog or very low cloud forms. Anticyclonic weather can then be extremely unpleasant and may continue like this for some time.

The properties of a depression, which we have just discussed, make it a very useful medium for weather forecasting. That is to say, they make it so if only we can foresee which way it is going to travel and how quickly; whether it will deepen further or fill and whether the fohn will set in and for how long. For the depressions travel about over the earth's surface, and the latter is not at all uniform, nor is it uniformly heated. The result of this is that before a depression we do not always get the same conditions existing and developing with mathematical regularity; nor is a depression transferred forward in a regular manner. Secondary depressions form and they, again, have the properties of any vortex (on a smaller scale). In this way the character of the overall weather pattern (the 'regional' weather) is obliterated, and local weather anomalies arise. Individual swaths of warm air (and associated swaths of cloud) can run ahead of a large depression, or accompany it in a lateral position and circulate around it. They then distort the weather picture of the large-scale depression, and all this has an effect on the forthcoming weather. These events are difficult to explain. Once they are recognized, however, they are a great help in forecasting the weather for a few hours ahead. Constant observation (at as many suitable places as possible), together with extremely rapid communications and the drawing up of weather charts, is, consequently, an essential prerequisite for a practical weather service.

Up to now, we have concentrated principally on depressions. Anticyclones are simply the reverse of these phenomena. Between two depressions there must always be a region of higher pressure, the formation of which is organically bound up with the low pressure vortices.

But there are also large anticyclones which are very stable, which do not move with the depressions, but which, nevertheless, have a marked influence on the tracks taken by the depressions. They have the distinguishing feature of extending far up into the troposphere, while the air near the surface is usually relatively warm (the so-called horse latitude anticyclones, like the Azores high). However, blocking anticyclones often form over the continent in winter and although the air in them is generally warmer than at the same level elsewhere, the temperatures near the surface may be very low because of the intense radiational cooling of the ground which takes place under clear skies. If, on the whole, there is less mention of these, it is because, to some extent, they do not figure prominently when our weather is being considered. They are always bound up with good weather — though we must also, of course, allow for those stable anticyclonic weather situations in winter which give rise to fog or lifted fog for days on end in central districts of Switzerland. In Britain 'anticyclonic gloom' is common in winter and the Scottish 'haar' which drifts off the North sea is also significant. At heights of more than 800 and 1200 metres above sea level, it is sunny and mild. Everyone regards this as the natural state of affairs, while everything else is felt to be a 'disturbance', at least in Switzerland. It is not that a high pressure area of this kind is the result of fine weather, or that fine weather is a result of high pressure. Once this kind of 'fine-weather-high-pressure-area' has formed it cannot be readily 'upset'; it stands like a wall and deflects oceanic depressions as though it were a large mountain. The weather within it 'holds out', and often does so despite distinct signs of poor weather.

If it were always 'fine', hardly anyone would comment on the weather; and if there were always a regular succession of good weather and poor weather, it would hardly be worth concerning ourselves with it very much.

The great thing about our weather is its changeability, its alternations and its ever-varying manifestations. Since, however, the weather is all-pervading, we naturally feel the desire (and often the necessity!) to know in advance what it is going to do.

As we have seen, most changes in the weather are, to some extent, an effect of the low pressure centre, except for instances such as fog clearing to give sunny skies without a low pressure affecting the area at all, and thunderstorms which may be solely due to the high surface temperatures achieved during a heat wave. If we want to know what the weather is going to be like, we must first of all know the extent of the depressions whether they are deepening or filling, how quickly they are moving and, what tracks they will follow. Once the track of a depression has been observed and recorded over an interval of time, it is often possible to conjecture what its subsequent path will be. Of course, this does not by any means always work out. With a few exceptions, the speed of displacement decreases towards the land. The majority of depressions which affect the weather in our part of the world originate in the eastern part of the south-east of the USA, from where they cross the north Atlantic to reach north-west Europe. But they can come from almost any direction. It is simply that it is relatively uncommon for this to happen.

The path followed by a depression is influenced by general conditions in the upper layers of the atmosphere. We do, however, have a great deal of information about these. We can say from statistical data received from constant upper air radiosonde ascents, that depressions favour certain well-defined tracks.

Only forceful individuals and mischief-makers are in the habit of not keeping to the ways of the general run of people; they bring disturbance and strife. So do depressions! Whenever one of them takes to a new and unusual direction, no one knows what the consequences will be.

But how do the well-behaved depressions move? First of all, they avoid obstructions and resistance. They go round mountains and avoid the 'stable' areas of high pressure. They also leave strongly heated regions to their right.

Without going any further into the 'why's and wherefore's', we shall now have a word about the five principal tracks — when each particular one is followed and what kind of weather (which 'weather type') we usually get from it.

All five are the continuation of long tracks which have originated much further off to the north-west or south-west. Of course, we want to consider the tracks of the depressions when they reach Europe. Track 1 (see Fig. 7) begins, as far as we are concerned, at the northern tip of Scotland and runs north-eastwards along the coast of Norway, over the Arctic Circle, there-

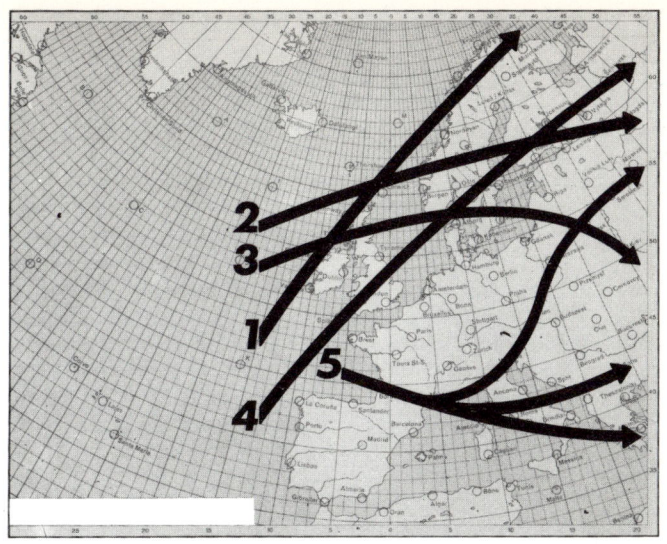

Fig. 7 Preferred tracks of depressions

after dividing into a number of secondary tracks which are of no further concern to us. The depressions show a liking for this track particularly in autumn and winter. The kind of weather which results from this track depends on the position of the nearest anticyclone. If its centre is over central Europe, then the oceanic winds do not penetrate very far in our direction, and in summer — a season when the depressions only follow this track on rare occasions! — we get very warm weather, whereas in winter it is very cold. If the centre of the anticyclone lies over the Alps, or even further to the south, the moist winds can penetrate a long way eastwards, and then strong south-westerly winds and rain can be expected, providing the fohn does not suddenly alter the sequence of events for an indefinite period (this can happen, too, with depressions on tracks 2, 3 and 4).

Track 2 takes the depressions from northern Scotland in an eastward direction over southern Norway and central Sweden towards Finland. This track is most frequently followed in winter. The depression passes nearer to central Europe than in the first case. As a result, stronger winds come, and more precipitation (this applies to central Europe as far south as the Alps).

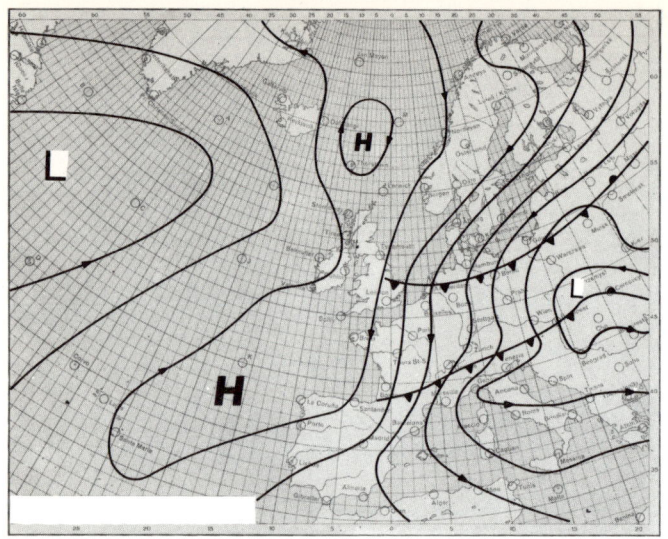

Fig. 8 Surface chart   H = High
                                       L = Low (depression)

Track 3 runs from the north of Scotland in a south-eastward direction, over the Skaggerak, across Poland and on into the south of Russia. This track too, is mainly followed in the winter. Depressions moving this way bring a lot of rain, thick cloud and generally gusty winds.
Track 4 is first picked up to the south-west of England. From there it runs to Denmark, then more or less over the Baltic Sea to Finland, and thence to the White Sea. This one could almost be called the usual summer track. Depressions following this track bring rapid changes of weather to the whole of central Europe, and they are usually (in summer) accompanied by strong heating and frequent thunderstorms. In the exceptional case of a depression taking this path in winter, severe westerly storms and north-westerly storms nearly always follow. They are often the cause of destructive tidal surges along the entire North Sea coast.
While depressions on tracks 1 and 4 bring rain and thaw in winter, those on tracks 2 and 3 produce snowfalls over low ground, especially if they had previously passed over Greenland. They become especially intense if a *Stau* situation evolves up against the Alps.

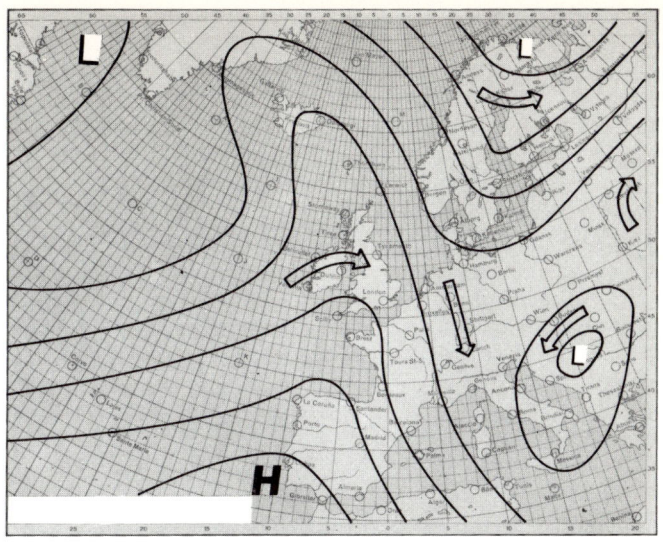

Fig. 9 Upper air chart (5500 metres)   H = High
                                                                        L = Low

Track 5 is the most interesting but least frequently occurring one. It occurs almost exclusively in winter. It travels from England across France to the Mediterranean, where it absorbs the depression over northern Italy which has come across the Mediterranean from the south-west, and then divides into three branches. One of the tracks curves away to the south-east, passing over Greece, the second goes eastwards to the Black Sea, and the third eventually passes over the area around Vienna in the northward direction. This third track is a very important one, even though it is only very rarely traversed (mainly in autumn or spring), because a depression coming this way generally brings unusually large amounts of precipitation to the districts which are affected (the southern side of the Alps).

These five tracks, known to us from experience, are by no means completely 'fixed' paths. A particular one is followed because that is where a depression finds the most favourable conditions for its advance. Since depressions are also affected by regions of high pressure and strong heating (as explained above), they have to follow a compromise track in many cases. High pressure and heating ('stationary maxima' or 'stationary highs')

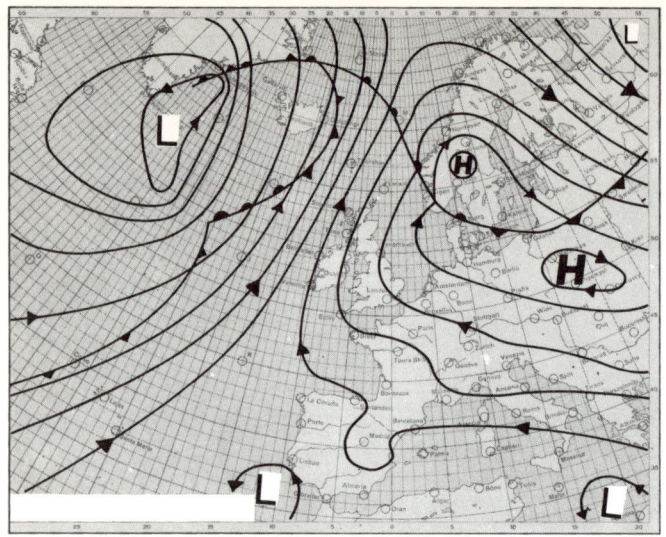

Fig. 10 Surface chart   H = High
                                    L = Low

are generally very stable conditions in respect of position and persistence. As a result, the depressions often follow the same track for as short as a few days and as long as a few weeks on end, because they have to keep avoiding the stationary highs. Thus, we get a succession of the forward side and rear side weather of the depressions for weeks on end. The weather has settled down; to being unsettled!

An anticyclone is situated over western Europe and a depression over eastern Europe *(see Figs. 8 and 9)*

The northward extension of the Azores high, which is also clearly in evidence at 5500 metres, pushes out over England and Iceland as far as Greenland. It links up at the surface with a northerly outbreak of cold air to form a separate high centre, but this is only at the surface. The ridge of high pressure to the west of the continent keeps all the Atlantic disturbances well away. All along its eastern flank, polar air (assumed to be moist) drives far to the south, piling up at first against the Alps. The isolated cold fronts, or stages in the cold air, are associated with a depression centred in the east of the continent and moving from north to

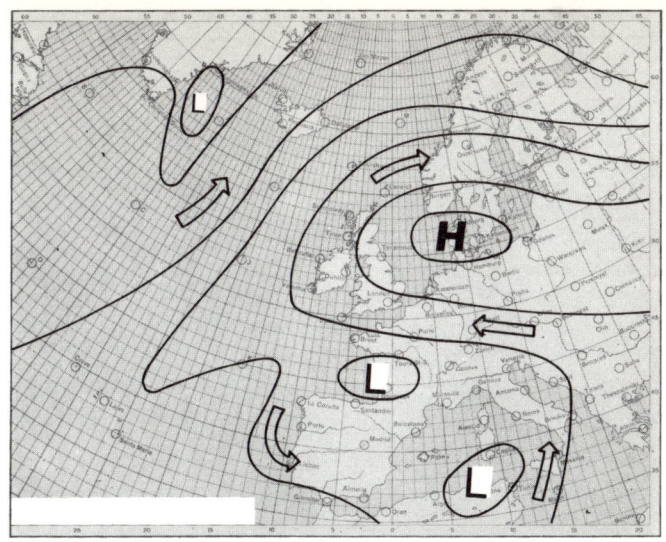

Fig. 11 Upper air chart (5500 metres)   H = High
                                                                      L = Low

south. The season is still only in mid-spring, so that snow can still be expected during the night down to 800 metres (in western Switzerland, only down to 1200 metres). On two successive days, 30 to 80 mm of precipitation fall in central Switzerland, but only about 20 mm in western Switzerland. In Ticino, too, in spite of a northerly fohn, perceptible cooling can be perceived. Yet there are only a few clouds about. The winds blow from the north at all heights. The next day it can be seen that the air, assumed to be cold, has a depth of only 3000 to 4000 metres. Above it the air is clear and dry, just as we find over the low ground 1 to 2 days later. The fine weather, which the incursion of cold polar air has preceded, should be fairly settled. It will, of course, be cool for a time with the arrival of the *bise,* in spite of the practically cloudless sky. In spring and summer especially, situations like this can produce a temporary return to cold weather accompanied by precipitation (the Ice Saints of mid-May).

An anticyclone is situated over north and north-west Europe with a depression over the Alps *(See Figs. 10 and 11)*

The upper air position of the high is to the north of Poland. There is a depression south-west of Portugal and one between the south of Italy and

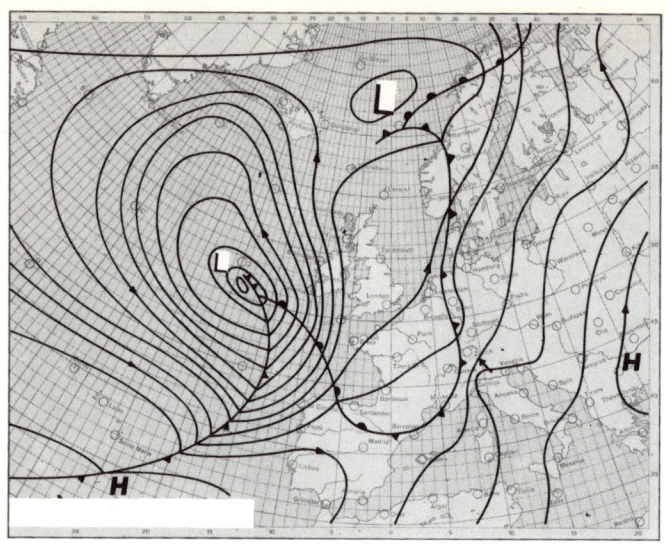

Fig. 12 Surface chart   H = High
                                          L = Low

Benghazi. At the upper level, however, the principal centres are displaced a little. The frontal troughs, which run from the deep depression over the northern Atlantic across Scandinavia and away to the Barents sea, no longer have any effect on our weather.

At the surface, and at the upper levels, too, the winds over Switzerland are blowing from a direction between east and south-east. Stations on the north side of the Alps are reporting a strong *bise*. As it gradually abates, the weather becomes settled. Since the season is already well advanced, this means the formation of widespread low cloud or lifted fog, (between late spring and early autumn it would be bright and dry, with cumulus and stratocumulus, even over the lowlands). It might not break up everywhere in the course of the day, but it will probably lift somewhat.

Peaks above 1200 or 1500 metres have bright weather and look out over a sea of cloud.

The southern side of the Alps, including Valais and Grisons, is not favoured either. In this region the presence of the low over the Mediterranean is making itself felt. There are reports of rain and upslope layered cloud. We

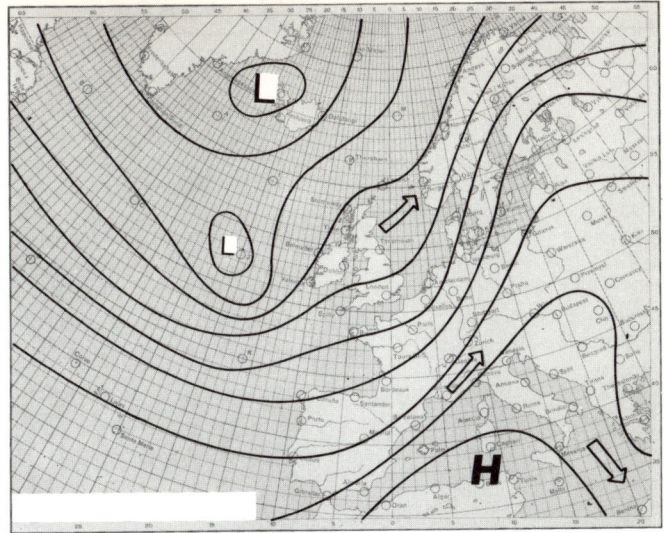

Fig. 13 Upper air chart (5500 metres)   H = High
                                                           L = Low

have here a limiting case between the two types of easterly wind situation. If the effect of the Mediterranean low does not extend to the Alps, the south side remains bright under the influence of the northerly fohn. If, however, the low is located in, say, the area round Genoa, the warm air from it is steered up against the Alpine ridge. The result is a *Stau* situation, with abundant precipitation on the southward-facing slopes. At the upper levels, however, the warm air slides away over the Alps to the north side, where it comes into contact with the cold air streaming in from the northeast. It undergoes intense cooling by conduction, which results in copious rainfall on the lee side of the obstruction as well.

An anticyclone is situated over eastern Europe with a depression to the west *(See Figs. 12 and 13)*

In this example, there is a strong upper ridge over Eastern Europe which is independent of the Azores high. Depression will travel eastwards across the Atlantic and then turn north-eastwards to Britain and Norway with the associated fronts weakening and becoming slow-moving as they move into Europe.

The winds are blowing over the Alps from a south-westerly direction causing an unrelenting fohn in Switzerland. It is also typical of the fohn that the isobars over the Alps on the surface chart have a so-called fohn bend' in them.

The 7.30 am observations from the climate stations in the fohn valleys are of interest in this connection. It can readily be seen that, even though it is still early March, they are reporting higher early morning temperatures (10 to 14 degrees C), in these valleys, with strong winds here and there, than in the low-lying districts (6 to 8 degrees C). The relative humidities are around 40 percent — on a March morning shortly after sunrise!

The weather chart also shows a cold front held up along the French Alps and the Jura. It will be midday before it gets into Switzerland. It is already raining in western Switzerland, and central Switzerland is completely cloudy. In addition, poor weather is getting in from the south up as far as the ridge of the Alps.

The weather which accompanies the fohn cannot go on being sunny and warm indefinitely. With regard to the wind, though, the southerly component continues. The fohn breaks down for the time being, but soon revives again — in advance of the approaching new front. It is possible that the cold air will not even manage to get into the valleys of the Alps and its foothills, but will have to be limited to invading the central areas only.

PARTICULAR WEATHER PHENOMENA

Water in the atmosphere

If there were no atmosphere, there would be no weather. This was why we introduced physical ideas about the air right at the beginning. We know that it is a mixture of various gases which are present in fixed ratios and that, in addition, there is a variable proportion of water in gaseous form. Biology teaches us that water is a fundamental prerequisite for all forms of life, and so it is for meteorology, too. Most weather phenomena are connected with the water in the atmosphere in some way or other. For this reason, we shall make a few more general remarks concerning the nature of water, before turning to weather phenomena in detail.

Chemically, pure water consists of molecules which are made up of one oxygen atom (O) and two hydrogen atoms (H). Its formula is therefore H_2O. In practice, however, chemically pure water is not found anywhere, because water is a most efficient solvent. It always serves in nature as a vehicle for all kinds of salts or acids, and for both inorganic and organic substances. Only rain-water is comparatively pure but, even so, not completely so. At the very least, it will have dissolved in it a little carbon dioxide from the air, not to mention waste gases and dust particles.

Since water is one of the most frequently occurring substances, it has often been used as a standard. Thus, a litre is the volume occupied by a mass of exactly 1 kilogram of pure water at its temperature of maximum density and under standard atmospheric pressure. Another way of expressing this is to say that the density, or specific mass, of water is 1000 kg/m^3 (equivalent to 1 g/cm^3). Even when measuring a quantity of heat, the unit employed until recently (the calorie) was based on water, one calorie (1 cal) being the quantity of heat required to raise the temperature of 1 gram of water by 1 degree C (more precisely, from 14.5 to 15.5 degrees C).

From what has been said, we might well expect that water is a substance which, compared with other compounds, possesses a very large number of 'normal' properties of the kind shown by other substances. This is, however, not the case. On the contrary, it has several unexpected peculiarities which, on close inspection, actually turn out to be exceptional properties, having most important consequences in nature. Among these, for example, is the peculiar variation of the density of water with temperature. Normally, the density of a substance increases as the temperature goes down. Solid substances and liquids contract when it gets colder, and the pressure of gases is reduced. Water behaves in this way, too, but only down to a temperature of +4°C. If the temperature falls any lower, the water begins to expand again, so that a cubic centimetre of it weighs less. At 0°C the water freezes to ice, with a sudden and much larger increase in expansion. Ice is lighter than water, whatever the temperature. This simply means that water which has not yet frozen sinks to the bottom, while ice always floats on the top. If water did not have this remarkable property, but behaved in a 'normal way', there would be a very different state of affairs. Ice would first start forming at the bottom of the body of water and any warmer water would move upwards. There would be no surface covering of ice to provide thermal insulation, and the seas would freeze through and through. Summer sunshine would not be able to thaw them. Aquatic life would no longer be possible.

We saw earlier that the gaseous water carried in the air represents a store of potential energy which can be released again, once the water vapour condenses back to liquid water. In this way, energy is transported over wide areas of the earth's surface. Though we may not see the sun in winter for days on end, we nevertheless receive some of its energy. The heat which is released when condensation occurs is being replaced way out there over the ocean through the evaporation of water under the influence of the sun. Heat is not only released during condensation ie: when water changes from the gaseous to the liquid state: it also occurs during freezing, when the water changes from the liquid to the solid state. Many areas of the earth's surface are habitable only because of this circumstance. In addition, there is the fact that, compared with other substances, water has a very high latent heat of vaporization. Hence, it turns out to be very well suited to its role in nature from the physical point of view as well.

By analogy with the name 'meteors' (those solid objects from outer space which fall through the atmosphere), we call all the forms of water which condense out of the atmosphere 'hydrometeors'. As we shall see further on, these include ice — snow- and water-clouds, fog, mist, dew, hoar frost, rime, rain, snow, ice pellets and hail and, finally, that far from estimable manifestation, glazed frost. In addition, there are many phenomena which can be indirectly traced back to those just mentioned. These include the evening and morning reddening of the sky, rainbows, optical phenomena in the atmosphere such as rings, haloes, and mock suns, and finally, shadow phenomena such as the Brocken Spectre. Mention should also be made of several manifestations, which have nothing to do with the water in the atmosphere, but are caused, it is often suggested but not proven, by electrons and ions in the upper reaches of the atmosphere; examples are the northern lights, and noctilucent clouds.

As we saw when discussing humidity, the process by which the water in the air becomes visible is always bound up with the cooling of the air in relation to the degree of saturation. Air can actually be super-saturated with water vapour to a certain extent. The presence of so-called 'condensation nuclei' plays an important part in the formation of clouds. These are tiny particles of dust or ions on which water vapour can be deposited in liquid form. We have all observed this process in connection with high-flying aircraft.

It often happens that we can see not the slightest trace of upper cloud, and then, all at once, our eye is caught by a high-flying aircraft which is leaving a condensation trail behind. The hot exhaust gases from the engines contain large numbers of ions. These are atoms which have a greater or smaller number of electrons than in their neutral state. The ions, together with extra water vapours produced by the engine, act as condensation nuclei for the instantaneous and localised formation of cloud. Under certain circumstances these 'condensation trails' can persists for a very long time.

Clouds

We shall now attempt to bring a little order into the multiplicity of types of cloud. An explanation of how clouds form and how it then rains has already been given early in this book. This explained the various types of cloud to be seen.

Note that not all clouds are composed of water or ice. In desert regions, there can be sand or dust clouds, which may then be carried further afield by very strong winds. Volcanic eruptions or immense forest fires fling smoke and ash far up into the stratosphere.

Cloud types have been carefully classified and given Latin names. There are ten genera (basic characteristic forms):

1 Cirrus	Ci	Feathery clouds consisting of ice crystals and having a fibrous or silky appearance
2 Cirrocumulus	Cc	Small, bright globules of cloud (small dapples) usually in widely extended areas
3 Cirrostratus	Cs	A thin veil of cloud
4 Altocumulus	Ac	Large dapples at a lower height, generally consisting of water droplets
5 Altostratus	As	Uniform layer cloud; usually thin enough to reveal the sun or moon, at least vaguely. It can consist of water droplets or ice particles, snowflakes or drops of rain can fall from it
6 Nimbostratus	Ns	Thick, dark, layered cloud, usually accompanied by rain. If the precipitation does not reach the ground, it can often be seen in the form of 'trailing precipitation'. Just below the Ns there are frequently low, ragged clouds
7 Stratocumulus	Sc	Layers of cumuliform cloud looking as though separate globules of cloud have been pushed together
8 Stratus	St	A low, featureless layer of cloud, often without precipitation
9 Cumulus	Cu	Heaped cloud; occurs in many varieties ranging from individual clouds of fair weather to towering thunderclouds
10 Cumulonimbus	Cb	Fully developed clouds with rain falling from them, sometimes accompanied by thunder and lightning

This classification largely corresponds to the height at which the cloud occurs. All cirriform cloud occurs at the higher altitudes, at the upper boundary of the troposphere, and is composed of ice particles. Altocumulus, altostratus and nimbostratus are medium-level clouds, while Ns, Sc, St and Cu are low-level clouds. Cb can extend vertically right through the whole troposphere; when the top of the thundercloud frays out into the form of an anvil, it has reached the upper limit of the troposphere and has already frozen to ice particles.

Besides these ten basic forms there are also many secondary species. For example, before the anvil-formation comes about, the powerful, swelling thundercloud is called cumulus congestus (Cu con). Altocumulus-castellanus clouds (Ac cas) have a crenellated appearance, and when there is a marked fall of temperature with height (ie a strong lapse), the turrets project upwards and are a sign of thunder. Lenticular clouds (lent) have a lens-shaped appearance, and indicate wave motion in the atmosphere. They frequently occur during fohn weather.

Clouds are often associated with fascinating optical phenomena. The lenticular clouds just mentioned often have an iridescent outline due to the diffraction of light at their edges. One can often see coloured haloes or coronae in the milk-white haze of Cs, as well as circles of light around the sun or moon (solar or lunar haloes). The smaller rings, called coronae, are due to the diffraction of the light by the water droplets. If they are coloured, the red edge is on the outside. They can also arise from diffraction by ice particles, in which case the ring shines with a dull white light. The larger rings around the sun or moon, on the other hand, are caused by refraction and reflection in ice crystals. They have an apparent radius of $22°$ or $46°$, and, if they are coloured, the red edge is on the inside. Other halo phenomena include the less frequently observed mock suns and sun pillars. Mock suns and solar haloes occasionally appear at the same time (see Fig. 14).

Experience shows that cirrostratus frequently occurs on the forward side of approaching depressions. Haloes are, therefore, fairly reliable signs of poor weather. The popular dappled clouds are frequently a sign of poor weather, too. There is something charming and cosy about the way they move across the sky in well-ordered and peaceful droves, but their appearance is deceptive. If they approach from the west or south-west, they are usually followed by poor weather. It is only when they approach from the east that it is not going to be so bad, since depressions from that direction never bring any serious deterioration in the weather.

Small, fair-weather cumulus clouds, which can be seen to dissolve into the blue sky of a summer's day, are a reliable sign of fine weather; so, too, are the large cumulus clouds which evaporate and disappear towards evening. If the development of the cloud has progressed too far, ie if it has become cumulonimbus towards evening, there may then be an evening thunderstorm, but this will not generally interrupt the continuation of the fine weather on the following day. How does it come about that fair-weather cumulus can build up to such enormous heights? A simplified explanation is as follows. As we have seen, when dry air rises in the atmosphere it cools by about $1°C$ per 100 metres and moist air by only about $½°C$ per 100 metres. If strong emission of radiation causes the decrease of temperature

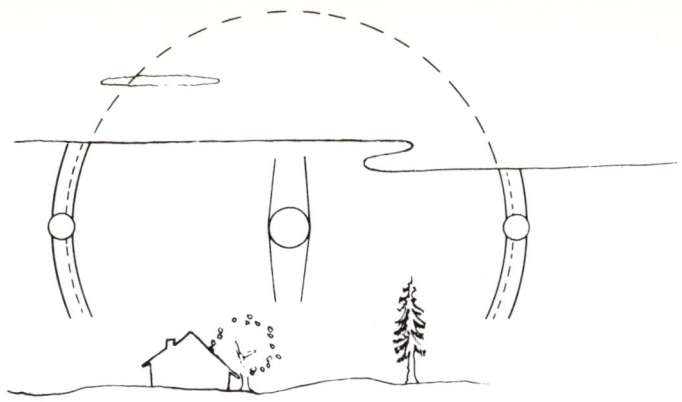

Fig. 14 Optical phenomena on 20.5.71 as observed near Lake Constance. In the high-level cloud, sun pillars, parts of the halo, mock suns can be seen

with height in the free atmosphere (called the lapse rate) to exceed these values, then an ascending volume of air will become more buoyant; at the point where the cloud begins to form, the buoyancy becomes even more marked, since there is less cooling within the cloud itself. The cloud is thus able to grow upwards at an increased velocity. This goes on until it reaches a layer of air in which the lapse rate is less than the net adiabatic lapse rate. It is in this layer that the anvil forms and spreads out. The formation of towering clouds takes place over favourable parts of the earth's surface, for example, those parts where there is variable-heating, such as mountain slopes which are particularly well exposed to the sun.

Fog

'Brother to the cloud', as the poet calls it. And rightly so! Fog is a large cloud lying on the surface of the earth, while fog patches are simply smaller clouds. Apart from its position, fog is in no way different from cloud. Just as with cloud, fog forms by condensation of the moisture in the air — when warm moist air moves over a strongly cooled surface and thus itself undergoes cooling. In temperate latitudes, especially after a long period of cold weather, fog of this kind occurs, almost without exception, when a thaw sets in. There are notorious fogs which form when the moist, warm air accompanying the Gulf Stream flows over the cold, northerly, ocean currents or comes into the vicinity of large icebergs.

Fogs of this type are extraordinarily dense and usually most extensive. In this respect, they differ from another type of fog which forms in the following suggested way: if the moist surface of the earth, or of a stretch of water is warmer than the air above it, evaporation of the warmer water produces vapour which soon saturates the overlying air and then condenses out as fog. The local fogs which occur in summer, immediately after a brief heavy downpour of rain, are of this type. They should be distinguished from the early morning and evening fogs of autumn. These occur mainly during anticyclonic weather when the air is still. The earth's surface loses heat by the emission of thermal radiation and, as a result, the lowest layers of the air are cooled down to their dew point. After fog has formed, there is further cooling by radiation from its upper surface, which is usually marked by an inversion (ie an increase of temperature with height) in the thermally stratified atmosphere. We can now understand why the morning sun disperses and prevents any further thickening of the fog and, indeed, it makes the fog evaporate. Wet fog, Scotch mist, drizzle and rain are just members of a long series. This situation is characteristic of a warm front.

Every droplet that forms requires a 'nucleus' to start it off. These 'condensation' nuclei are the tiniest particles of dust, soot and smoke. With the great increase in the production of such nuclei in many places (such as London, Hamburg, the Rhineland between Basel and Frankfurt) in the last century, the incidence of fog also increased to an enormous extent. Now we have already learned that the air at great heights is as good as dust-free. Yet clouds still form there, and even small local patches of fog. The condensation nuclei in these cases are the ions which are always being produced under the influence of solar radiation, especially in the upper layers of the atmosphere.

Haze and mist

'Haze' (humidity below 95%) and 'mist' (humidity above 95%) are the names given to turbidity in the atmosphere which is so slight that the sun can still cast noticeable shadows. The turbidity can be due to actual pollution produced by finely divided substances of all kinds (volcanic ash, dust raised from sandy terrain, smoke). This is 'dry haze' or 'dust haze'. It is responsible for the most striking and splendid chromatic effects due to the refraction of light, and for the most colourful sunsets and twilight phenomena.

It is the moist turbidity, or 'mist', however, which occurs much more frequently. It is an important constituent of the air and, by its very nature, it is just a limiting case in the formation of fog and cloud. As with fog and cloud, it consists of the tiniest droplets. It differs from the other two, however, in that it is spread out through the air over huge distances. In winter its upper surface is very much more sharply defined. Frequently the mist lies in the form of a second layer above a 'sea of cloud'.

If, as a result of either thickening or thinning out, mist and cloud gradually merge into one another, it is not easy to distinguish a boundary. The point of transition has been arbitrarily defined as the place at which the horizontal visibility becomes one kilometre. If one can see further than that, the turbidity is called 'mist' or 'haze'; if less it is called fog.

Visibility in air depends on the amount of mist or haze in it together with possibly rain, falling snow or sea spray — and not only its transparency, but also its (apparent!) colour. The blue of the air is a result of the scattering of white light by the molecules of air. Since blue is the most strongly scattered part of the spectrum, it comes to us from all parts of the sky. In the absence of mist or haze, the sky is pure blue. In the complete absence of any air, on the other hand, it is black — and it does indeed tend to be nearly black over the highest of the earth's mountain peaks. When the particles of mist or haze suspended in the air are large, they reflect the white light of the sun, and the sky then appears to have a white, 'pallid' hue.

It can be readily appreciated that the formation of mist or haze is encouraged by proximity to the earth. The upper level of the so-called 'haze layer' is at about 2000 to 3000 metres in summer, and rather lower in winter. The top of the haze layer marks the transition between the layer in which air has recently been in contact with the ground and the 'free atmosphere' which is relatively undisturbed. This also explains why it is the autumn and winter months that bring the 'view of the Alps' from the mountains of Southern Germany, apart from a few exceptional days in other seaons when the air is particularly pure and, therefore, clear. In advance of, or 'underneath', upward gliding warm air, the atmosphere has an upward motion. As a result, the atmosphere seems 'clean' — hence its remarkable clarity. It is true that particularly clear, 'pin-sharp', distance views are not usually a good weather omen. This is especially so in summer when, as we have said, it is a matter of experience that the haze layer reaches up to a greater height than in winter.

The mysterious charm of landscape with deep relief is to be ascribed exclusively to haze. It is haze that creates distance and 'visible' air; it is the painter's 'atmosphere'. In lands where there is no haze, almost everything seems to be equally far away. It is the lack of haze which is responsible for the unfamiliar, strange impression we have of many exotic landscapes (eg Tibet, Bolivia, Arabia).

The optical effects we see when shafts of sunlight become visible in the atmosphere are also due to mist or haze. They are, however, of no value for forecasting purposes. When light from the sun, shining through gaps in the clouds, encounters particles of water or dust floating in the air and illuminates them, long bright beams appear. Adjacent to these are the darker bands of air which have not been made visible.

Red skies in the evening and morning

Atmospheric turbidity is also involved in the twilight effects of the evening and morning sky. With completely clean air, the sky would slowly become paler (the intensity of the scattered light decreasing), and the area immediately around the sun would become a yellowish red, since all the short wave light is scattered on its long path through the atmosphere.

Since turbidity, of whatever kind, is much more concentrated in the lower layers of the atmosphere than in the upper layers, the rays from the setting or rising sun have to travel a long way through the turbid layers. Because most of the blue light is scattered out on the way, red light predominates. This is how the 'red sunset' and 'red dawn' comes about. On days with fine, clear sunsets, we see all the gradations of colour in the western sky from the brightest yellow to the darkest scarlet. At the same time, the twilight in the eastern sky spreads out with less luminous hues (mainly a dull ochre and a gloomy purple). Overhead, the sky is still a rich, deep blue, which merges through shades of bluish-green, green and yellowish-green into the colours of the western and eastern skies. This twilight, and the colours in the sky which accompany it, come to an end when the sun has sunk $18°$ below the horizon. We can see from this why the twilight in the tropics only lasts a short time. It is because the sun goes down so 'steeply'! It is a particularly splendid spectacle when isolated clouds in the sky are illuminated, partly by light shining through them and partly by light shining onto them from below.

Red dawns and red sunsets give useful indications about the weather to come. A delicate red in the evening, with pure hues, tells us that for a long way to the west there are no clouds in the sky and that for the time being there is no depression approaching. When clouds, which are the precursors of a coming disturbance, are already very thick in the west, there cannot be any display of colours.

Apart from a pure enjoyment of nature, we do not like seeing a bright red dawn! And for a good reason! The coloured sunrises which attract our attention only occur with isolated clouds and haze in the east. Disturbed weather may then be not far away, in which case the western sky will soon be showing signs of it!

Assuming that disturbances usually follow a track from west to east, the popular saying is quite right: "Red sky at night, shepherd's delight; red sky in the morning, shepherd's warning".

Rain

'Rain' is the name given to water droplets which fall out of the air onto the earth. It is produced, like other forms of precipitation, by the condensation of vapour when the air is cooled below its saturation point. We have already mentioned the causes of cooling, viz. mixing with colder air (which produces only light falls of rain), radiation (which mainly results in fog or dew) and expansion of the air without there being any exchange of heat with the surroundings (ie 'Adiabatic' expansion). This expansion can only take place during the ascent of air. It results, first of all, in the formation of minute droplets which are held in suspension by the ascending current of air. In fog, the diameter of a droplet is about 0.01 millimetre, but in very wet fog it can be up to 0.1 millimetre. In a cloud which is developing upwards at a rapid rate, (and, of course, in this respect is really nothing other than fog), the drop size in extreme cases can increase up to 7 millimetres in diameter. If the drops are heavy enough to fall down to the ground, they are then called rain. This happens when the upward force, exerted by the ascending current of air, is no longer strong enough to counteract the downward force of gravity. The size of the drops is, therefore, a function of the velocity of the ascending air within the cloud from which the rain is falling, and by their collecting together as they fall at different speeds because of their size. In order to hold drops of about 5 millimetres in suspension, the upward current of air must have a speed of at least 7 metres per second. The speed at which the drops come down does not much depend on the height from which they fall since, as a result of air resistance, the speed of descent soon becomes uniform. The amount of rain which has fallen is expressed by scientists as the depth of water which would cover the ground if none of it flowed away. How much rain (or, rather, water) can a given volume of air hold? This is entirely dependent on its temperature. One cubic metre of saturated air, when cooled by one degree C, would produce 0.2 grams of water at $-10^{\circ}C$, 0.6 grams at $+10^{\circ}C$ and 1.6 grams at $+30^{\circ}C$. If a column of air, 1 square meter in cross section at ground level, 1000 metres high and saturated at $20^{\circ}C$, were raised through a further height of 1000 metres, it would produce 4 kilograms of rain (which is equivalent to 4 millimetres of precipitation on the earth's surface). The amount of rain in a single cloud is, therefore, by no means limitless. It is a well known fact that very frequently heavy rain becomes lighter and lighter, the longer it continues raining. However, if it keeps on raining 'steadily' — which occurs mainly with westerly winds — it means that new streams of air are flowing into, and rising up over, the continent. The rain will then continue as long as this type of weather persists. It is, naturally, the edges of great mountain barriers which are most susceptible to this kind of effect. With depressions which are not excessively deep it may well happen, for example, that the weather in the Alps is better than in the immediately surrounding belts. The most intense rainfalls which have been observed are about 5 millimetres per minute, but they only last a short time.

Rainwater (condensed vapour) is quite pure in itself. While falling through the air, however, it carries along with it all kinds of substances which are present in the atmosphere, principally, dust (dirty drizzle), soot (black rain), pollen (sometimes yellow rain), and desert dust from the Sahara when there is a powerful, warm, southerly airstream (orange-yellow rain, or snow in the Alps). In addition, rain dissolves a whole series of salts and acids which are always present in the air. It is these impurities which make rain water dangerous to buildings and monuments made of limestone and marble. They are also responsible for the patina which forms a coating on copper roofs in the course of time. But it is these very substances which are also good fertilisers and justify the farmer's saying: "A lot of snow manures the soil". Naturally, the largest amounts of rain must fall at places where new clouds are always being formed. This will happen where warm, moist currents of air are continually being forced to rise. In Europe, this occurs first at the western edges of the continent and then on the western sides of the mountains.

As a result, the amounts of precipitation in neighbouring areas can differ a great deal. All those places which are in the lee of a mountain, and so are within its 'rain shadow' (ie in the shadow of the prevailing wind), do not get very much rain, while those on the windward side have abundant rainfall. The driest places are, of course, those large valleys, surrounded by mountains, where the wind from any direction must move downwards to penetrate into them. It is conditions like these which account for the abundant sunshine in, for example, many Alpine valleys in the cantons of Grisons and Valais, qualifying them as health resorts.

The distribution of rainfall over large areas (whole continents) is shown on rainfall charts, where lines of equal amounts of precipitation (called 'isohyets') are drawn. From these charts we can deduce that the largest amount of precipitation in Europe falls in the southern part of Dalmatia: the value is about 4,500 millimetres.

In the absence of general and persistent precipitation, the favoured time for rain is the warmest part of the afternoon, because of the upward motion of the heated air. The morning hours are preferred by fogs because it is cooler and therefore provides better conditions for condensation. There is a good rule, based on experience, that the weather can be expected to clear up (if it is going to happen at all!) between 9 and 10 o'clock in the morning, and that if there is any tendency to precipitation, the preference is for 2 to 3 o'clock in the afternoon (3 to 5 o.clock in summer).

Rainbows

Usually, the observer only sees one part of the bow; and he only sees this part providing the sun — while he is looking toward the area where the rain is — is behind him. It is only when the sun touches the horizon that the rainbow becomes a complete semicircle, the centre of which lies on the extension of the line joining the sun to the eye of the observer. The observer has to be between the sun and the place where the rain is falling in order to see the ring of colours which lie next to each other. On the outside of the ring at the upper edge, it is red, while at the inner or lower edge it is violet. Between these two there is the series of colours orange, yellow, green, blue and indigo. Sometimes, on the outside, there can be a second adjacent rainbow which is less luminous, and with its colours in the opposite order. Finally, but very infrequently, there can be second order rainbows lying on the inside of the primary bow, so that there may appear to be up to six altogether.

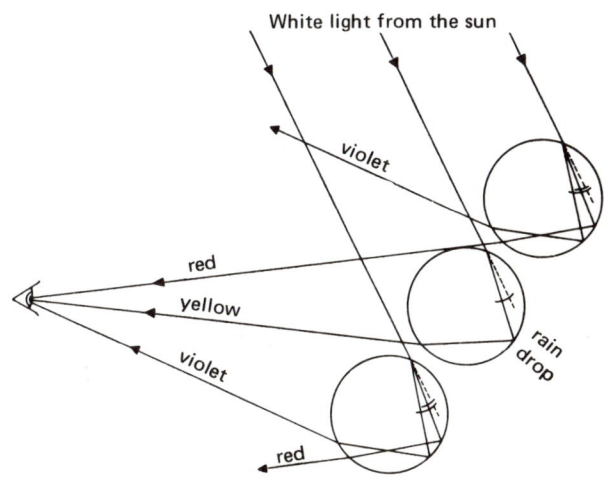

Fig. 15 How the rainbow is produced

The primary rainbow is produced by refraction of the rays of light as they enter and leave each raindrop; in addition they undergo reflection at the inner surface of the drop (see Fig. 15). Upon refraction the white sunlight is split up. As a result, only violet rays arrive at our eye from the lower raindrops, the other colours, blue through to red, passing by below our angle of view. That zone, therefore, appears to us as violet. Red light enters our eye only from the raindrops in the upper zone, the violet, blue and yellow rays passing by above our angle of view; hence droplets in the upper zone shine with a red light. Here we have an application of that law of wave theory which says that, when light passes into or out of a denser medium, the various wave lengths, or colours, are refracted to varying degrees, the long waves (red light) less so than the short waves (blue light). All the differences between rainbows in respect of width, colour sequences, and brightness are due to the size of the raindrops. A luminius pinky-violet together with a bright green indicates drops of 1 to 2 millimetres in diameter. Drops of 0.5 millimetre in diameter produce only green and violet, with no yellow (in the second order bows). If there is yellow in the inner bow, while at the same time the primary bow is wide and brightly coloured, we can assume that the drop size is 0.2 to 0.3 millimetre; and so on. A rainbow visible in the east means that there must be clear air to the west, because this is where the sun is shining from; and since our prevailing winds are westerlies, so that our 'weather' comes mainly from the west, there is a good prospect that we shall be getting bright weather. On the other hand, a rainbow in the west tells us that there is rain there, so this rain is very likely to come our way too. If, however, a rainbow appears in association with the isolated showers of otherwise fine, summer weather, then, naturally, this rule of thumb does not apply.

Snow, ice pellets and hail

Snow, ice pellets and hail are three different stages of the same process, which starts with the formation of snow and finishes with the formation of hail.

When a column of air expands and cools, and the temperature is below $0°C$, the water vapour in it changes directly to ice crystals ('snow crystals') without first becoming liquid. This change, directly from vapour to solid, is called 'sublimation'. The colder the air, the smaller the snow crystals which are formed, corresponding to the smaller moisture content. The nearer the temperature of the air approaches freezing point (or even exceeds it), the more the crystals tend to aggregate to form flakes. The largest flakes observed up to now have a diameter of up to about 10 centimetres. If snow crystals combine with supercooled water droplets, tiny opaque balls of snow are formed which can be easily crushed and which are called ice pellets or soft hail. If a further layer of ice is deposited on the little, round ice pellets during their long fall through the air, they are then called hail. This means that hail always comes down from a great height! If hailstones are carried aloft again in powerful, ascending currents of air, they stick together in lumps. This is how those giant hailstones are formed which sometimes grow into great lumps of ice of irregular shape and great weight.

Snow

The word 'snow' is used for both the falling flakes and for the covering they produce when lying on the ground. There is as endless a variety of types of lying snow as there is of shapes of snow crystals. If snow is compressed, the individual crystals melt or, at least, become so soft that they become severely deformed and fuse together. At very low temperatures, the ice crystals fracture under pressure and the snow 'crunches' under the wheels of a car or the soles of our shoes. The depth of the snow cover is generally much less than the layman is inclined to suppose. Where the wind piles up the snow, naturally quite enormous snowdrifts can form here and there, but these tell us nothing about the amount of snow which has fallen. A snow depth of 30 centimetres is seldom exceeded over open level country. In mountainous districts, of course, a single fall of snow can attain vastly greater depths. In the Alps, a snowfall of 1 metre or more in 24 hours has frequently been observed. Even here, though, the snow does not accumulate to the great depths mentioned in fairy tales, since it packs down under its own weight. It is said to become firm (old snow which is in the process of becoming compacted into ice), and the overall depth of snow would be unlikely ever to exceed an upper limit of 3 metres.

The ratio of the depth of water which would result from the melting of snow over a given area is called the 'water equivalent' of the snow. For freshly fallen snow, the ratio is about 1 : 10; in other words, a 10 centimetre depth of snow produces about 1 centimetre of water. Snow, especially when freshly fallen, reflects both visible light (which is why it dazzles us so much) and ultraviolet light (which is why our skin becomes such a deep brown so quickly). The snow surface emits strongly in the infra-red and therefore undergoes intense cooling. Being a good reflector of sunlight it heats up a little during the day. However, since the snow is a poor conductor of heat (due to bubbles of air) the loss of heat at the surface cannot be made good by transferring heat up from the ground. Therefore the ground under the snow remains near freezing while the top surface may be many degrees centigrade lower. Snow thus helps to moderate the effect of severe air frosts. Hence, it has an important influence on the weather conditions in general.

Small hail

Small hail, or ice pellets, has a diameter of 2 to 5 millimetres and represents the transition between snow and hail. Usually, it only comes down in small quantities Over the plains it occurs mainly in the spring when frost is light and the weather is unstable and squally. It is also more frequent by day than by night. It is preceded by snow or hail, but only rarely followed by them. In the mountains, falls of soft hail, or ice pellets, occur fairly often in the summer, too, and nearly always accompany thunderstorms.

Hail

We have already seen how hailstones are formed from ice pellets by the deposition of additional spherical layers of ice, and then how they produce giant hailstones by further agglomeration. The general size of hailstones is between that of a pin and a orange. Stories of very large lumps of ice should be accepted with considerable reserve. It is quite easy for hailstones to freeze together on the ground, producing enormous pieces of ice, but they have not fallen from the clouds in this form.

Hail is associated with strong, upward currents of air which carry the raindrops to heights where the temperature is well below the freezing point. This is how completely, or almost completely, transparent stones are formed. In this case the hail is really a frozen cloudburst, and we can readily understand why such a large amount of ice falls all at once.

A hailstorm without a thunderstorm is an extremely rare event because they both result from the same cause, viz., a sudden and localised ascent of the air over a restricted area ('local instability'). Moreover, since thermal thunderstorms — as we shall see — are of only limited size, though their tracks often extend over long distances, and since they also prefer to travel along certain paths, it is fairly obvious that hail behaves in the same way; 'the hail comes down in zones'.

Sometimes hail is preceded by isolated large drops of rain; these are hailstones which have melted. The rapid cooling of the air by these masses of ice is the reason why the hailstones which follow do not also melt and become raindrops. From what we have said about hail and the way it is formed, we can readily understand why hail is a phenomenon which is more likely to happen in the warmer season, and in the warmer part of the day.

Dew

Strictly speaking, dew is the same as the moisture on a glass of cold beer which has been brought into a warm room. Grass and leaves are very good conductors of heat and cool very rapidly by the emission of radiation which, of course, can only occur at night when the sky is cloudless (a twin sheet of cirrus does not alter the long wave cooling substantially). They become considerably colder than the surrounding air, and exactly the same thing then occurs as we saw with the glass of beer. The temperature of the air in direct contact with each blade of grass, or leaf, or the glass, falls to the dew point and condensation then occurs in the form of drops of water. We can readily see that any movement of the air (wind) which mixes the layers of air with each other, acts against the formation of dew, or even completely prevents it. Moreover, it is clear that the dew will be far more abundant after prolonged cooling through the night than when it starts to separate out the previous evening. Naturally, the more moisture the air contains, the more dew is formed. This is why the formation of dew is chiefly a summer phenomenon. It is also why the formation of dew in the tropics is so much more abundant than in our part of the world.

Dew, then, does not 'fall', but is deposited. In this case, too, every droplet needs its condensation nucleus. A night and early morning with abundant dew would not amount in our country to more than 1/3 millimetre of precipitation. In the tropics, under certain conditions, the quantity is ten times as much!

Hoar frost

The formation of hoar frost follows the same laws as those applying to dew, but instead of condensation we have sublimation. Hoar frost consists of very tiny ice crystals, the size of which depend only on the temperature of the air and the amount of water vapour it contains. The colder the air, and the less the moisture in it, the smaller are the ice crystals. As soon as the dew point falls below zero degrees, hoar frost forms, and it would really be more correct to call this temperature the 'hoar frost point'. Hoar frost also has a preference for precipitating onto the individual flakes of snow lying on the ground. The arrangement of the individual ice crystals is often in the form of rows, and it produces feathery structures which are entirely composed of crystals.

Rime

If relatively warm, moist air arrives after a prolonged cold spell, and if it is cooled to below its dew point (which happens especially when it passes over extensive patches of cold snow), the water vapour separates out as a fog which, to begin with, is composed of drops of supercooled water. When these supercooled droplets strike against a solid object they immediately freeze onto it and cover it with ice, mainly, of course, on the windward side. If this occurence alternates with that of wet snow, which is impacted against trees, poles and rocks, already covered with ice, to adhere there and freeze (and which then in turn receive another coating of rime), the shells of ice and snow can build up to such an extent that, under certain conditions, they can become a metre thick and attain an incredible weight. Telegraph poles turn into thick, white trees, wires acquire a diameter of 20 to 30 centimetres (before they snap!) and bushes turn into fairy-tale shapes. The whole landscape becomes enchanted. A typical form of rime in the mountains and foothills of the Alps is one in which the ice is in a compact granular or amorphous mass.

Hoar frost and rime are two very different things! The first is a delicate tracery, weighing almost nothing, and hardly able to bend a blade of grass to the ground. The other is a coating of ice weighing scores of kilograms, transforming trees and shrubs into ponderous structures. And yet, in origin, they are closely related.

Drizzle, granular snow

These should be mentioned for the sake of completeness. Drizzle, which also goes under other names in some districts, consists of fine droplets of rain having diameters less than 0.5 mm. Granular snow consists of fine grains of snow having diameters less than 1 mm. Produced only at low temperatures, it comes about in just the same way as drizzle, but in solid form.

Glazed frost

This is the meteorological name for a smooth, clear, transparent coating of ice which forms on the ground, on trees, on fences, etc. It can form in two different ways. The first way is by fine rain, or even a warm, moist airstream striking against cold objects well below zero degrees C. The second way is by rain falling through a layer of air which is at a sub-zero temperature, and thus becoming thoroughly cooled itself to a temperature below freezing point without, however, immediately solidifying. Upon impact against a solid object it instantly turns into transparent ice.

In the latter case (which, however, only occurs rarely) quite incredibly large masses of ice can accumulate which result in great damage, especially in forests.

Thunderstorms

Thunderstorms are always associated with powerful upward currents of air. These rising currents are produced when moist air near the surface of the earth becomes more strongly heated than its surroundings. This depends most of all on the nature of the terrain: open fields, woods or water. Flat, open country and sloping ground emit quite different amounts of heat into the air lying over them. Once the ascending currents of air have started, they intensify spontaneously for various reasons (as we shall see), until the development ends in a thunderstorm. Thunderstorms formed in this way are called 'heat thunderstorms'. The conditions for the formation of a thunderstorm are not always to be found in the non-uniform heating of the terrain. An ideal situation is, more frequently, the warm front of a depression, where a warm air mass comes up against a cold one and slides up over it. These are called frontal thunderstorms.

While the ascending current is concentrated into one or more, usually fairly narrow, channels, the air subsequently flows down towards the surface over a wider area. The air in the ascending current rises because it is warmer than its environment. If this relation between the temperature of the ascending air and that of the environment changes during the course of the ascent in such a way that the temperature of the surrounding air at a great height is warmer than that of the ascending air, then we have a stable stratification in the atmosphere, and this acts like a brake on the upward motion of the ascending current. If the stratification is unstable, the ascending current of air will always be warmer than its environment, even though it is cooling as a result of the decrease of pressure with height; hence, it will continue to experience an upward force. The clouds thus produced act, at first, as a load on the rising currents of air which have acquire additional buoyancy. As soon as rain begins to fall, however, the upward velocity becomes greater than ever. Initially, the raindrops do not fall right down to the surface, but evaporate into the oncoming, upward current of warm air, making it still moister. As a result, it reaches its dewpoint even more quickly.

As can be seen from the schematic cross section of a thunderstorm in Fig. 16, a well-developed towering thunderstorm reaches as far as the upper boundary of the troposphere. The air which is shooting upwards inside it cools to below zero degrees C. The water droplets do not freeze at once. They can cool down to as low a temperature as $-10^{\circ}C$. Just as there must be nuclei present for condensation to occur, so there must be freezing nuclei for there to be freezing. These are provided by dust particles in the atmosphere which are always present in sufficient numbers. Their physical properties determine the temperature at which they become effective as freezing nuclei. The reason that the frozen water droplets falling towards the surface do not form pretty snow crystals, but large hailstones, is to be found in the large upward velocities associated with the processes going on in the towering thunderstorm.

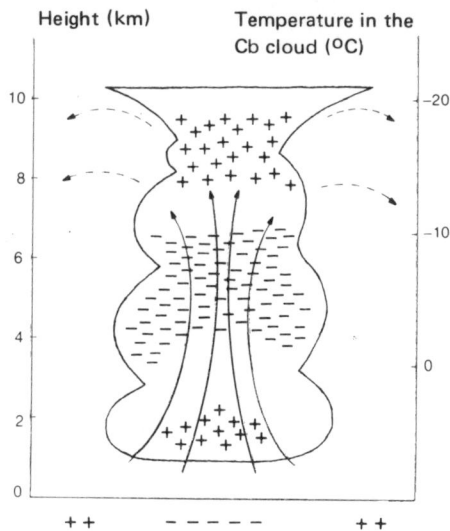

Fig. 16 Cross section through a thundercloud

A towering thunderstorm — *cumulonimbus* — is charged at the condensation stage by the latent heat.

The ascending current of air carries the charge, (actually a positive one), and distributes it within the cloud. The falling particles of ice carry a negative charge downwards.

As we can see in the diagram (Fig. 16), the inner part of the cloud is negatively charged, while the upper part is positive. Positive charges can be identified in the lowest section of the cloud, too. Owing to electrical induction, there are corresponding negative charges on the earth's surface. As the charge builds up, the potential between the individual parts of the cloud and the earth's surface also increases. The air is a good insulator for electric field strengths of up to 10,000 volt/cm. The potentials created in the thunderstorm, however, go up to the order of 100 million volts. Whenever the spatial distribution of charges results in the field strength (ie the potentials related to their distance apart in centimetres) exceeding the value mentioned above, we are in the regime of discharge phenomena. At 30,000 volt/cm the discharge is in the form of sparks. Thunderclouds and lightning have been thoroughly investigated by various kinds of research methods. Although there is still some doubt about certain details, we have a very good idea about the way in which lightning takes place from investigations using rotating photographic apparatus. When a flash of lightning takes place between cloud and earth, an electrically conducting path is first created by a preliminary discharge. This preliminary discharge usually takes place from cloud to earth. It travels in a stepwise fashion at a rate of 10 kilometres in about 0.1 second, and is only faintly luminous. The strength of the associated current can be as much as 1000 amperes. Once the preliminary discharge reaches the earth's surface, the main discharge commences, and a current, a hundred times as strong, travels from the surface up along the prepared channel to the cloud. The picture of lightning takes place between cloud and earth, an electrically conducted take the word 'lightning' to mean the very much brighter main discharge, then the process occurs only very rarely. Frequently, the same channel is used for several discharges, giving one the impression of longer flickering, but in this case also the eye is unable to perceive the individual events. This impact of the accelerated charge carriers against the air molecules causes intense heating of the air, and this produces a pressure wave which is audible as a crack.

We see the light emitted by the lightning discharge practically instantaneously since light propagates at a speed of 300,000 km/sec. The sound, however, only travels out at 330 m/sec. This means that, not only do we hear the thunder from a distant flash of lightning 3 seconds later for every additional kilometre it is away, but that, since strokes of lightning can be several kilometres long, under certain conditions, we hear not just a single crack but a prolonged rumbling and rolling.

The type of lightning described above is called 'forked lightning'; it is the most common type. Streak lightning occurs between the earth and the cloud or between separate parts of clouds.

In addition, there are a few rarer forms of lightning such as 'pearl-necklace lightning' and 'sheet lightning'. Up to now, there has been no satisfactory explanation for these. Brush discharge lightning and St. Elmo's fire are glow discharges. In conclusion, heat lightning, is not in a category of its own but is really very distant forked lightning, so far beyond the horizon that we only see it as a diffuse light, and its thunder is no longer audible.

The development of summertime heat thunderstorms has already been explained. The series begins with small fair weather cumulus, which disperses in the evening. It then goes through the larger, swelling clouds of the following days until it gets to the mature cumulonimbus, whose thundery showers bring a redeeming coolness on sultry, summery days. Heat thunderstorms occur most frequently, not only in the hot season, but also in the hottest hours of the day. Many a cumulus has already swollen powerfully upwards, and has even taken on its typical anvil of delicate ice needles, but still without thunder breaking out, for night approaches and inhibits further development. Heat thunderstorms move and change their direction in what often appears to be a quite arbitrary fashion. Of course, the ascent and downward motion of the air involved in a thunderstorm do not take place in the beautifully symmetrical manner depicted in our schematic cross section (Fig. 16). If, as a result of the asymmetry, the thunderstorm has been displaced a little, the place which it has just left will be colder. This gives rise to a difference in pressure which pushes the thunderstorm still further on. The track of a heat thunderstorm is very dependent on the local conditions.

Something should be said about the danger from lightning. Fork lightning can smash roofs, tear down trees and set fire to wood. If it strikes a person, it leaves burns; frequently the victim dies. Iron objects which are near the path of the lightning become magnetised as a result of the magnetic field generated by the intense current in the discharge channel. In the vicinity of the discharge channel, molecules of oxygen combine to form ozone. This is what causes the 'stink of sulphur' which folklore has associated with lightning from olden times. Folklore also has a number of rules about how to behave during a thunderstorm, but these do not stand up to scientific examination. It is quite true that one should guard against seeking shelter under large trees from storm and rain, but which trees are safe and which are dangerous? Large isolated trees are dangerous because the electric field above them takes the form of a cone, and if a preliminary discharge occurs in the vicinity it is very likely to be 'drawn' to the top of this cone. Lying flat against the ground, however, can be just as dangerous as standing upright. If there is a lightning stroke nearby, a voltage gradient is produced around the point of impact, and anyone unfortunate enough to be lying in the direction of this point gets the full voltage gradient between his head and feet. Hence, it is safer to take up a crouching position.

People are always sceptical about new scientific facts which they cannot explain with the knowledge they acquired at school. This enables us to understand why, following an introduction to the practical advantages of electricity — usually well bolstered up with all kinds of superstitious knowledge — many people take up a defensive attitude towards this new suggestion, instead of quite simply studying the principles of electricity. Completely false ideas have still been kept alive. Obviously, a high tension transmission line lightning conductor is much more effective in attracting lightning than a tree. This is because the tree, being made of wood, even though it is water in the tree which is conductive, is poorly earthed, while the transmission line, on the other hand, is very well earthed. Hence, such a conductor is unquestionably a protection for the houses in its vicinity.

Aurora

The aurora, or northern lights, is a phenomenon which takes place in the ionosphere (see the cross section diagram in Fig. 1). The earth is surrounded by two Van Allen radiation belts, the existence of which is due to the earth's magnetic field. The first of these belts is situated at a distance above the earth's surface equal to half the radius of the earth; the second is two or three times the earth's radius away from the surface. They are both annular formations, something like an apple which has been hollowed out. Particle radiation from the sun (protons, electrons, 'the solar wind') gets trapped in them and is prevented from reaching the earth. Over the magnetic poles of the earth, where the radiation belts are at their weakest, particles repeatedly cascade down from radiation belts into the ionosphere where they collide with the molecules of air, thereby giving rise to the luminous phenomenon of the aurora.

THE WEATHER SERVICE AND WEATHER FORECASTING

An answer to the question "What is the weather going to do?" is not only desirable or of interest for many people, but is often of vital importance for air travel, or climbing mountains. As we have already shown, one must first of all observe the weather in order to ascertain the laws which govern it, and in order to be able — on the basis of these laws — to produce forecasts of its future development. Furthermore, in order to make a forecast it is of decisive importance to have a knowledge of as many measurements and other observational data as possible from that region of the atmosphere which is important for the forecast in question.

Because of the multiplicity of the events involved, the earth's atmosphere, together with the processes going on it, should be regarded more as a living organism and less as a physical contrivance. Although the individual processes are subject to the laws of cause and effect, there arises a fundamental question as to whether a phenomenon of such complexity can ever be accurately forecast at all; or whether we might not, in principle, have to be content with forecasting some — even though fewer — of the possible future events.

One thing is sure; the further the forecast reaches into the future, the less reliable it becomes. If, nowadays, forecasts for the next few hours are fairly reliable, then those for the next few days are distinctly more doubtful, while forecasts for the next few weeks or months cannot be made with any degree of ultimate reliability, although, in reality they would be of more value.

Though we perceive light, sound and temperature directly with our senses, we are still subject to various errors. In order to avoid such errors we make use of measuring instruments, and these also make accessible a range of effects which we cannot perceive directly with our senses: electricity, magnetism, invisible radiation. The instruments can also send us measurements from spatially distant regions which we cannot reach ourselves, but which are important to us in the further development of the weather. Radiosondes send us the pressure, temperature and humidity conditions in the free atmosphere, and, in addition, information about wind speed and direction at the upper levels. Weather satellites transmit television pictures to us from hundreds of kilometres above the earth, showing the distribution of clouds over whole continents.

The science of weather is quite a young blossom on the tree of knowledge. The reliable collection, recording and evaluation of weather reports and measurements is only a hundred years old. We should not be surprised, therefore, if meteorology cannot provide us with sensational successes in forecasts. One often hears that people who spend a lot of time in the open are able to make good forecasts purely intuitively or by the feel of the weather. When examined closely, this faculty turns out to have less to do with feeling, and much more with having a decidedly good gift of observation and, in addition, being able to think critically. Thus, such people have made a start on the science for themselves. This is quite sufficient for local, short-period forecasting but, unlike the individual observer, the weather service has the advantage of being able to view a larger area, and this increases the reliability of the forecast.

The usefulness of being able to view the whole collection of observations in this way depends entirely on the speed of communications. It is not surprising, therefore, that the beginning of systematic weather observation and, along with this, the beginning of modern meteorology, coincides with the introduction of the telegraph and telephone. The first telegraphic weather report appeared in 1848 in the London *Daily News*. During the Great London Exhibition of 1851, a daily weather chart appeared for the first time. It was based on observations from 22 weather stations. 1855 saw the appearance of the first weather chart in France. Finally, in 1904, the first radio weather report was given to the *Daily Telegraph* by a ship at sea.

Today, there are thousands of observation stations operating in Europe and the adjacent territories, transmitting complete weather reports every three hours round the clock. In the north Atlantic there are a few stationary weather ships, while large numbers of merchant ships also send in observations. Every country has one or more upper air stations which carry out radiosonde ascents. In addition, there are other stations which have been entrusted with special tasks (eg radiolocation of thunderstorms). Air traffic has a weather service of its own in which the exchange of weather reports takes place over shorter time intervals. The meteorological service for aviation obviously co-operates closely with the ordinary weather service. The weather reports of individual stations are transmitted by radio or land-line to the central offices of the countries in question. There, they are immediately processed and then re-transmitted at definite intervals to all other centres which require them. Nowadays, facsimile telegraphy is already in use, and even small ships may be fitted with easily-managed equipment which can receive weather charts by radio and print them out. As in the case of radio-astronomy, meteorology, too, experienced enormous growth during the last war. Not only were previously unknown methods of observation tried out for the first time in the war (eg observations with the aid of the radio-sonde) but, even more the whole science of meteorology benefitted from the enormous amounts of money which countries granted to their meteorologists in connection with strategic plans, troop transports and the war in the air.

The forecasts put out by the weather service are seldom for a period exceeding 24 hours. Forecasts can be useful for this kind of interval. At the same time, it should not be forgotten that the weather forecasts broadcast by radio stations in their news bulletins, by the telephone weather service and by newspapers, are all regional forecasts and only apply as an average for a large area. Supposing the forecast says' "Gradually becoming warmer, fine, tendency to thunderstorms in the afternoon with precipitation in places". This does not say much about the probability of a thunderstorm at a particular place. We only know that the tendency to heat thunderstorms is increasing, but whether the clouds will already be gathering during the morning and we shall get a heavy downpour of rain with thunder and lightning, entirely depends on local conditions. But it is of value if the individual is doing his own weather study because, combining the latter with the weather broadcast, he will soon be in a position to understand the peculiarities of the place where he lives, and to make correct forecasts. That is the only point, when all is said and done, of the weather forecasts broadcast by the news media. One cannot really expect that listening to a weather forecast with half an ear during lunch could have much practical advantage.

The sarcastic couplet: "When the cock crows on the dung heap, the weather'll change or stay as it is" has more truth to it than one might at first think. It means that at any given moment there is a fifty-fifty chance of making a correct forecast if one either asserts that the weather will be fine or prophesies that it will be poor. There is a much bigger chance of being right if one says that the weather will stay the same as it is, because many weather situations, whether we are thinking of, say, summertime anticyclones or of unsettled 'April weather', show a marked persistence. A forecast which is too general no longer has any value as information. On

Fig. 17 (Turn picture so that coat is at bottom). Weather satellite photograph. A typical depression near a coast. To the left and in the middle of the picture, the rain clouds of the cold front; to the right of this the warm sector with cirrus clouds; to the right and above, the warm front

Fig. 18 Weather satellite photograph. Right and below can be seen Ceylon with the mainland to the left. The streaky clouds on the right, if observed from the surface of the earth, would give a similar impression to the polar bands in Fig. 36. The clouds just to the right of the centre of the picture give clear evidence of having been produced by wave motion in the atmosphere

Fig. 19 Photograph looking towards the south-east just before sunrise. It is autumn, and a sea fog is covering central Switzerland right up as far as the mountains, against which it is starting to rise

Fig. 20 Stratus over the south side of the Alps. On the air route Zurich - Rome. Position: Verzascatal and Maggiatal. View towards the west, with Mont Blanc. Height of cloud: upper layer 2000 - 2300 m. Flying height about 8000 m. Date 15.4.1971 at 0800 hours

Fig. 21 Cirrus spissatus cumuloimbogenitus. Lacking a supply of moist air, a cumulonimbus has dissolved. This cirrus has been left over from its anvil

Fig. 22 Fog patch or cloud? Physically, they are both the same. The patch of fog is a cloud lying on the surface of the earth

Fig. 23 The 'sun drawing water' (the sun's rays, passing through an apperture in the clouds, and illuminating suspended particles), along with a yellow colour in the atmosphere, is a sign of high humidity

Fig. 24 The highlands of Sutherland (Scotland) between Lairg and Tongue, in the vicinity of the Crask. Stratocumulus

21

22

23

24

Fig. 25 Altocumulus lenticularis clouds; a fohn situation. Wilderswil, Interlaken. View towards the east

Fig. 26 Same place. View towards the south-south-east

Fig. 27 Summer cumulus cloud. While the clouds at the extreme left would still be recorded as the relatively well-behaved fair-weather type, those towards the middle of the picture, further inland, are rising up more vigorously. To the right of the middle of the picture small turret-shaped tops (altocumulus castellanus) can be seen. These indicate strong instability in the atmosphere. There will be thunderstorms

Fig. 28 Cumulonimbus calvus

Fig. 29 Schleswig Holstein. Cumulus congestus

Fig. 30 Cumulus cloud, building upwards strongly. In the background an anvil is forming

Fig. 31 After the passage of a thunderstorm. Neuhaus, Lake of Thun, Switzerland. View towards the west

Fig. 32 Low ragged clouds of bad weather (stratus) with fohn-like clearing up in the evening. The lens-shaped clouds with irridescent edges, often occur in association with the fohn

Fig. 33 A mackerel sky: pretty, but not a sign of good weather. The cloudlets are arranged in patterns which point in a definite direction

Fig. 34 Altocumulus floccus

25

26

27

28

29

30

31

32

33

34

Fig. 35 Picturesque cirrus clouds at sunset. The cirrus indicates that there is a depression not far away. A red sunset in a case like this guarantees that the weather will be fine for a while

Fig. 36 The sun is going down behind an increasing cover of dense cirrostratus. The bands enable us to tell the direction of the wind in the high troposphere. The fine weather has come to an end

Fig. 37 Rapidly increasing cirrus. In the background there are cumulus clouds, strung out in a line at a constant height

Fig. 38 We have learnt about weather situations which are typical of our climate. We know how a depression develops, with its warm front and cold front, what phenomena are associated with its front system, and how the passage of these fronts generally takes place. Weather charts nearly always look complicated, and the development of the weather does not do us the favour of keeping to a set pattern, but produces new variants.

Let us just take a look at the 0100 hours chart published in the weather report of the Swiss Central Meteorological Office, Zurich, for Good Friday, 9 April, 1971. We see that 7 low centres have been indicated in a fairly flat overall distribution of pressure, together with some frontal systems which may make a rather bewildering impression. There is a depression centred over Biscay with a warm front, a precipitation area, and a cold front, the latter going over into the frontal system of another depression centred over the southern part of the Mediterranean. The depression over Biscay poses a threat to the fine weather which there has been over Switzerland and southern Germany up to now, and poor weather might be expected. How is the weather going to develop?

35

36

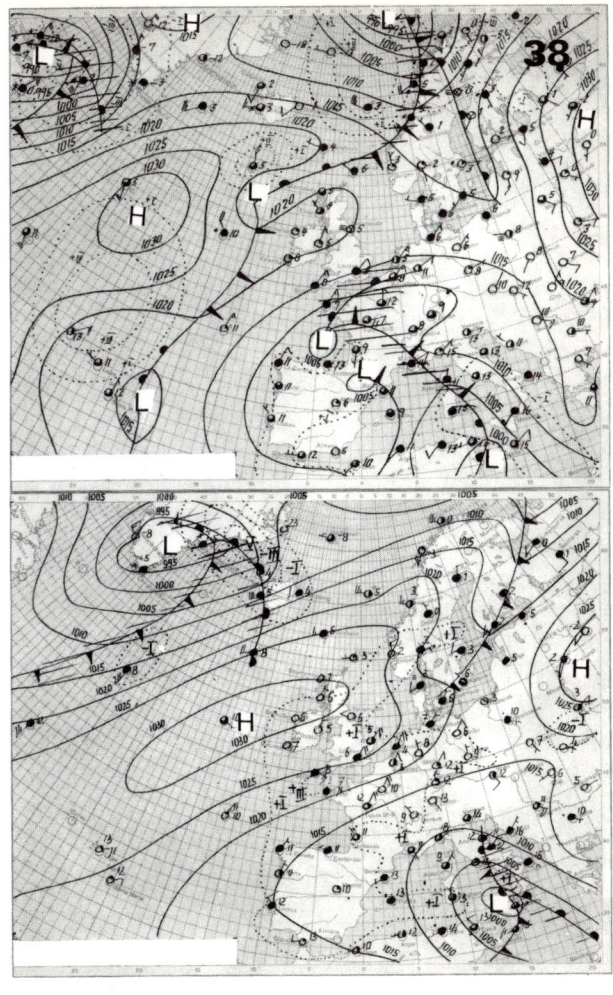

the other hand, one should on no account go into too much detail in a forecast but should limit oneself to forecasts to average values in time and space. The good weather forecaster will, therefore, keep to a middle course between these two extremes. For example, he will be as careful about expressions like 'violent thunderstorms', 'storm force winds' etc., as against stating extreme temperatures. According to the laws of probability theory, as the period for which the forecast applies becomes longer, its accuracy must diminish — and so it does! Anything going beyond a fairly detailed 24 hour forecast, or a forecast of the general type of weather over the next few days, is, at the present time, very uncertain. If a forecast is put out for a very cold winter or a rainy summer, it should be accepted with the greatest caution. This does not mean that the larger meteorological forecasting. Just as we can say today that a significant advance in weather institutes are not diligently carrying out research into long range observation and reporting took place with the introduction of modern means of communication, so we may one day say that weather forecasting made its decisive advance with the introduction of electronic computers. The masses of data which have to be considered for a long range forecast are far greater than a scientist can cope with using pencil, paper and a slide rule. Just collecting the data 'by hand' is a boring task which takes up a lot of time, and it can only be carried out on a satisfactory scale when it is taken over by machines. This new technique is called 'electronic data processing'.

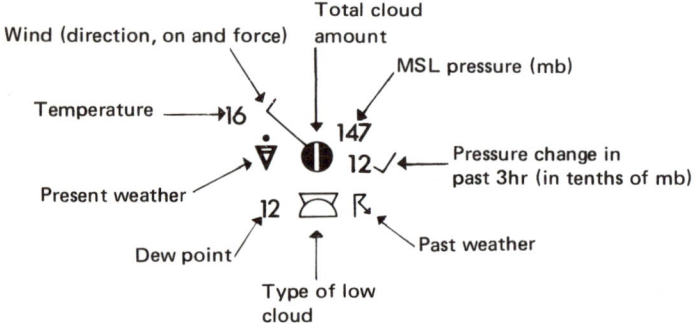

Fig. 39 The station model

There is another point. The physical laws of thermodynamics and fluid mechanics which govern the processes developing the atmosphere, are of staggering mathematical complexity for the layman, and even the specialist finds the work of carrying out the calculation of a problem presented by a given weather situation so time consuming that he would not want to take it on too frequently. In this respect, the electronic computer is certainly a long-awaited tool.

The method of coding the weather messages, enables the job to be done relatively quickly even without data processing machines. Once the weather messages from the various stations have arrived at the central office, the surface and upper air charts can then be prepared using the plotting symbols. For this purpose outline maps are used which cover an area from America to Russia and from the Arctic Ocean to North Africa. These charts are already printed with the weather stations in their correct positions. Once the chart has been plotted with the data reported from each station, according to station model, the first thing is to draw in the isobars. These are the lines joining together all places with the same air pressure. Since it is only rarely that two places report exactly the same pressure, the isobars are always drawn between neighbouring stations so as to interpolate the point where the pressure corresponds to the value of the isobar being drawn. The isobars are usually drawn at 4 millibar intervals. The temperatures and weather phenomena are examined against this background in order to find the fronts associated with the depressions. Finally the areas of precipitation are shaded in.

When the new weather charts are ready they are compared with the earlier ones. This enables us to gather what development has taken place, and making the forecast only involves extrapolating the development into the future. This sounds quite simple but is not so by any means. It is usually clear to some extent, of course, how the matter is going to proceed, but as to when the development will have arrived at any given point, this is usually much harder to say. To be sure, the meteorologist will always be making comparisons with similar earlier ones, but the development does not proceed exactly like 'it did before'. Some new aspect will always supervene, and this can prejudice our decision about the form and speed of the further progress of events.

WEATHER RULES

Weather rules are principles of weather development which have been derived from years and years of experience and which have usually been put into the form of little verses to make them more readily apparent. Nowadays, we often cannot avoid the feeling that some rules exist mainly to satisfy the rhyme and, actually, many traditional weather rules do not stand up to a scientific test.

Our weather is dominated by a constant seasonal rhythm. In summer it is warm and thunderstorms occur; in winter it is cold, thunderstorms are rare but there is fog and snow. Furthermore, the weather of the temperate zone is nearly always changeable, and long periods with the same weather only occur rarely. In the tropics and the polar regions, the seasonal changes of weather repeat themselves with a greater degree of regularity. In our latitudes, if we were briefly to forecast the weather for each day of the year as that which occurred on the corresponding day of the preceding year (ie based on a long term average) there would be more chance of this occuring than any other weather.

An early collection of relatively carefully compiled weather records, (in Switzerland), covering a seven year period, goes back to Dr. Mauritius Knauer, Abbot of the Langheim monastery near Kulmbach. His almanac went around in copies from hand to hand, and what originally started as a register of observations soon developed into credible forecasts. One may blame the circumstances of the time for the fact that Knauer tried to demonstrate that the seven heavenly bodies already known in antiquity, viz. Saturn, Jupiter, Mars, the Sun, Venus, Mercury and the Moon, took it in turns to determine the weather, and did so according to the properties ascribed to them by astrology.

Directly after the invention of printing, books began to appear about the weather, and weather omens. The time had come to put into print, and into the hands of educated people, the ancient wisdom of the farmers, which had more or less succumbed to astrology. It was in this way that Knauer's records, too, came out in 1701, being published under the title *A Hundred Year Almanac* by the doctor, Christoph Hellwig, of Thuringia, who was also a clever businessman. The hundred year almanac gave the weather to be expected for every day of the year, irrevocably fixed in the wisdom of astrology, and it became a real book of the people. When the almanac turned out to be correct, it was celebrated as a triumph, but when it was wrong, it was overlooked. In this regard we should remember that even a 'blind forecast', of the kind described, has a fairly high probability of being right.

Even today, popular belief has it that the moon has a special part to play in weather prophecy. This is not surprising, because, in addition to certain effects which the moon actually does exert on the earth, such as the ebb and flow of the tide, there are other effects, in part less real, on people and animals. It entices the moonstruck from his bed, cows are said to show a preference for having their calves on nights when there is a moon, and moonlight is said to be good for warts. Whatever the weather is at the moment, a change can be expected at full moon or at new moon, and anyone who disputes this will often make himself unpopular. This problem has been subjected to a very thorough scientific examination with the aid of statistics covering many years, and it turns out that there is no connection at all between, for example, incursions of cold or warm air and the phases of the moon.

This does not mean that the moon has no effect on the weather, but it does not happen in such an obvious way. The moon riases the tides, and it even moves the earth's crust a little, so why should it not affect the atmosphere? It is all a question of the intensity of the effect. For example, there is definitely a relationship between air pressure and the phase of the moon. In the tropics (at Singapore) it is possible to measure a periodic oscillation in air pressure of 0.06 mm of mercury, which is caused by the moon. In Europe, amplitudes of only 0.01 mm of mercury have been encountered. This is, of course, extremely small, and one cannot really imagine how such tiny oscillations of pressure could affect the weather. But this problem must be regarded as unsolved.

Scientists in the USA have looked for a relationship between the amount of daily precipitation and the phase of the moon, and they claim to have found a weak correlation. It is still being disputed, however, whether it is a sufficiently clear effect to be accepted as a palpable fact. Connections between rainfall and the moon could be explained on a scientific basis. Thus, it appears that dust suspended in the atmosphere promotes the formation and, hence, the precipitation of rain; regarded from the point of view of cybernetics, this is an extremely significant regulatory process. Like any other kind of matter, dust is subject to the law of gravitation, of the attraction between masses, and when we come to discuss the attraction of the moon, we are thinking less of the dust produced by volcanic activity or industry, and much more about the cosmic dust which falls onto the earth out of space in considerable quantities. This dust is not uniformly distributed everywhere in the universe. Along the paths followed by the meteor showers around the sun, there are clouds of dust and larger meteoric particles in an irregular distribution. This material gets into the earth's atmosphere whenever the earth's orbit intersects that of the meteor shower, which always happens at the same time of the year. That is why we are able to witness a regularly returning display of shooting stars, sometimes on a greater scale, sometimes on a lesser one. It is by no means fanciful, therefore, to suppose that the moon, because of its position and by way of gravitation, might affect the fall of this dust; and that this would show up in the precipitation which is affected by dust.

It is considerations like these which really land us in the middle of astrology. To be sure, cosmic dust does not appear as a factor in the horoscopes of the old astrologers, and if we find out, one day, that the planet Jupiter has some effect or other on us or on our weather, then this new knowledge will hardly be in harmony with that of the old astrologers. The thing that should certainly be discussed further in this connection is the influence of the sun. Certain trees in some localities reflect the 11 year sunspot period in their tree rings. This means that the trees have sometimes grown better and sometimes less well in rhythm with the frequency of sunspots. There is a suggestion, of course, that this effect occurs through the weather, but when we try to correlate particular weather phenomena with the sunspot period, it is not easy to find anything significant. Similarly, the water level in Lake Victoria is quite well correlated with the sunspot period, but the inevitable attempt to correlate the amount of rainfall in the catchment area of the lake with the sunspot period came to nothing. Somehow, we are not yet done with the complexities of weather phenomena. The pressure and wind in the upper atmosphere over the tropics show a measurable periodicity of 26 months. Up to now, however, we have simply not discovered what causes this period. There is still a great deal of research to be done.

Whereas astrology has tried to get behind the laws of the weather 'from the top' so to speak, but without a sufficient observational basis, the farmers, with their weather rules based on long and patient observation, have collected mainly practical knowledge. Sayings like: "When winter cold becomes less severe we shall soon see snow", "If dark clouds are about, you can say there will be rain", "A rainbow in the morning, shepherd's warning" and "Red sky at night, a shepherd's delight" are useful weather rules. It should not be forgotten, however, that many of these sayings only apply to the locality, and cannot be transferred to another place just as they are. Many rules attempt to say something about the fairly distant future on the assumption that the warm part and cold part of the year must compensate for each other. Such a compensation can, indeed, be verified on a statistical basis, but only over long periods of time. It is not true that a cold summer is necessarily followed by a mild winter.

It is not worth considering those sayings which try to infer the weather of a later month from the impression of a particular day. "If it is sunny on St. Urban's day the wine harvest will be good, take it from me; but if it rains it will turn out bad, as we know from long experience." (That is, whatever the weather on St. Urban's day — 25th May — this is what the general nature of the weather will be for the rest of the year!) The rule of the Seven Sleepers belongs to this category, too, promising that Seven Sleepers day (27th June) decides the weather for a period of seven weeks. In spite of rules like these seeming very like superstition, they have been examined statistically, but, as expected, with no positive result.

Many popular weather omens also involve an increase in humidity. If we ask a shepherd for his opinion about the weather to come, he will probably put on a mysterious expression and feel inside the fleece of one of his sheep. That makes sense! For hair of all kinds, including, in its way, horny skin, has a very strong attraction for moisture ('hygroscopic'); it absorbs water, becomes soft, and then no longer rustles. Of course, the shepherd would hardly know about that. He is simply aware that when the hair of his animals is in that sort of condition, 'there is rain in the air'. It is known that there are quite a few animals which give very reliable warnings about the weather. It is less well known, however, that they are nearly all influenced by the moisture in the air. One example is the flight of the swallows. If they fly high, it indicates good weather, while if they fly close to the ground it foretells poor weather. We can also point to 'fishes gasping for breath' before heavy rain, to the mounds of earth thrown up by moles, and to spiders resting. These are all signs that poor weather is on the way. Quite right: if we look into the matter, they are all signs that the air is already humid.

The fact is, that the swallow flies where it can find most food. If the insects are lower down, close to the ground, the swallow flies low, too. Fishes have no need to gasp for air; it is just that they are leaping after flies which are buzzing about immediately above the water. The mole notices at once when insects hide away in the ground, and it is then active close below the surface, with the result that it throws up a higher mound. The spider stops working on its web when so much prey has been caught in the snare that it can hardly catch up with wrapping them up and killing them — the insects have come down from above and crawl about on the ground as soon as the air becomes humid; this is due to an instinct for self-preservation which has developed through countless generations — at higher levels they are caught up by the wind, and the forceful beating of the rain drags them down. Swallows, fish and moles (and many other creatures, too) probably know nothing about what kind of weather is on the way — and yet we can draw conclusions from their behaviour.

The effect of humidity is also responsible for the tiny movements made by certain 'weather prophet' plants. Cell walls or stalks swell up owing to the humidity in the air, and this results in turning or bending or even closing of the flower. Rules which refer to the appearance of cirrus clouds are usually correct. If such clouds are approaching from the west, we may confidently infer a change in the weather, since they run ahead of the area of low pressure, often by hundreds of kilometres. The more quickly they move, the sooner we shall get the poor weather. If they approach from the south-west, we may expect the centre of the depression to pass over us and the unpleasant weather to be upon us very shortly. In certain areas in the foothills of the Alps, cirrus clouds are not a reliable indication that poor weather is on the way, showing once again that weather rules are very dependant on the locality! In this case the fohn is going to set in, and in certain circumstances it will not be clearing away from the area very quickly. Cirrus coming from the east, on the other hand, is more of a sign of continuing bright weather.

By using rules of this kind for forecasting — and even scientifically trained observers do so — and introducing some sort of measuring instrument into the observation, we arrive at the modern form of weather rules. In conclusion, therefore, we shall give 12 rules. It is assumed that a small aneroid barometer is used.

1 If the air pressure goes up sharply (by 4 to 6 mm) over a period of a few hours, the bright weather will only be short-lived.

2 If the pressure rises considerably during the course of a day, fine weather can be expected, and its duration will be in proportion to the rise. If the pressure rises for only one day, then the good weather will not last much longer than that.

3 If the rise is slow, uniform and prolonged (two or more days), a longer period of dry weather is in prospect. If, on the other hand, the wind veers from west towards north, it will soon clear up; (in autumn and winter there will be stratus or lifted fog in the valleys).

4 When there is a pronounced rise in pressure, an improvement in the weather is to be expected, particularly when the wind, having been in the south and then gone to west, veers further until it comes from the north-east.

5 If the barometer goes up to an unusually high value during a calm and the humidity is high, fog can be expected to form, but it will often be followed by bright weather.

6 If the pressure is rising abruptly and in jerks, but falling a little in between, this usually indicates the arrival of unsettled weather; likewise, when it is falling abruptly and in jerks, with brief rises in between.

7 When the pressure is falling, one can confidently expect precipitation, if, at the same time, the wind goes round from north or east to south or south-west; always providing that the fohn does not intervene!

8 Prolonged and persistent falls point to continuous precipitation. The longer the fall goes on, the longer the precipitation will last. If the fall is unusually abrupt (and deep), there will be strong winds in addition to the precipitation.

9 During a calm, if the fall, though abrupt, is not deep, and the weather is very warm (especially with increasing humidity in summer), then a thunderstorm can be expected.

10 The early arrival of rain can be more confidently expected if the fall occurs between 10.30 and 11.30 in the morning, and continues. With westerly winds, the rain will usually start within the next 24 hours; with easterly winds it will be a little later.

11 If the pressure goes up only in the afternoon, even if only a little, there will usually be an improvement, though it may not last for long.

12 If the pressure falls in the afternoon, but only a little, this does not mean much, especially in the summer. This afternoon fall has to do with the 'diurnal variation of pressure', and is only due to the air being warmed.